NANOTECHNOLOGY SCIENCE AND TECHNOLOGY

NANOWIRES

SYNTHESIS, ELECTRICAL PROPERTIES AND USES IN BIOLOGICAL SYSTEMS

NANOTECHNOLOGY SCIENCE AND TECHNOLOGY

Additional books in this series can be found on Nova's website under the Series tab.

Additional e-books in this series can be found on Nova's website under the e-book tab.

NANOTECHNOLOGY SCIENCE AND TECHNOLOGY

NANOWIRES

SYNTHESIS, ELECTRICAL PROPERTIES AND USES IN BIOLOGICAL SYSTEMS

LUKE J. WILSON
EDITOR

nova publishers
New York

For permission to use material from this book please contact us:
Telephone 631-231-7269; Fax 631-231-8175
Web Site: http://www.novapublishers.com

LIBRARY OF CONGRESS CATALOGING-IN-PUBLICATION DATA

ISBN: 978-1-63117-855-9

Published by Nova Science Publishers, Inc. † New York

CONTENTS

PREFACE

Zinc Oxide (ZnO) is a metal oxide semiconductor of group II-IV whose nature resides at the borderline between covalent and ionic semiconductors. In the last decade it has raised several attention in the research field since it possesses promising catalytic, electrical, electronic and optical properties. It can be easily prepared in different shapes and sizes at both the micrometric and nanometric scale, including the forms of micro- and nanowires, nanobelts, nanotubes, micro- and nanoparticles, multipods, tetrapods, and flower-like microstructures. This different variety of morphologies has thus attracted considerable attention for potential application in solar cells, nanogenerators, field effect transistors, gas sensors and other electronic micro- and nanodevices. This book discusses the synthesis and the electrical properties of ZnO nanowires. It also discusses rough silver nanowires, nanobuds and nanoparticle substrates; the synthesis and properties of Ni nanowires in porous silicon templates; and phonon scattering and elastic energy propagation in nanowires.

In Chapter 1 the authors aim to overview the synthesis techniques and growth mechanism of zinc oxide (ZnO) nanowires and their related properties. In particular, here they review both wet and dry synthesis techniques reported in the literature, such as hydrothermal and sol-gel routes, template-assisted growth, electrodeposition process, chemical vapor deposition (CVD), pulsed laser deposition and RF magnetron sputtering.

According to each synthesis technique, the chapter reports not only the detailed procedure, but also the advantages and disadvantages, analyzing in particular the properties of ZnO nanowires related to their microstructural and morphological features. The functional properties of ZnO nanostructures are also overviewed with a particular attention to their electrical properties and the related potential applications, exploiting either the semiconductive and piezoelectric behavior of ZnO nanowires. A comprehensive review is here focused on their use as field effect transistor, energy nanogenerators, pH and gas sensors, as well as the use of ZnO nanowires for energy conversion, like photovoltaic cells and water splitting photoelectrochemical cells.

Chapter 2 introduces an improved solid-state ionics method combined with vacuum thermal evaporation to synthesize silver nanowire, nanobud and nanoparticle structures. Surface enhancement effects in the silver nanostructures are studied, and provide a reliable approach towards the trace analysis of biomacromolecules. First, the authors introduce the solid-state ionics method for synthesizing silver nanostructures, and discuss morphological changes resulting from different growth currents. Second, they discuss surface plasmon polaritons excited on the silver nanowires. Third, the authors detect the surface-enhanced Raman scattering signal of low concentrations of Rhodamine 6G on silver nanostructure substrates. This demonstrates that the rough silver nanostructure surface provides a significant surface enhancement. Fourth, the effects of natural oxidation on the surface enhancement of the silver nanostructures are studied. The effect of oxidation on the silver nanostructure substrates, such as their thickness, surface roughness and particles size, is discussed. Oxidation strongly affects the surface enhancement of the silver nanostructures. Finally, the nanostructures are applied in the surface-enhanced detection of λ-DNA. The Raman signal of a single λ-DNA molecule is detected, as a result of the surface enhancement by the silver nanostructured substrate. The enhanced fluorescent signal of λ-DNA is detected at more than 10 μm from the substrate surface. The rough silver nanostructures exhibit excellent surface-enhanced Raman scattering, and provide a new approach for biomedical research.

In Chapter 3, structural and magnetic properties of Ni nanowires (NWs) electrochemically deposited into pores of porous silicon (PS) template under the stationary galvanostatic regime have been investigated. Samples have been studied in detail by using scanning electron microscopy (SEM), X-ray diffraction (XRD), Raman spectroscopy and magnetic measurements. SEM analysis revealed the formation of PS/Ni nanocomposite at the initial stages of Ni deposition with the characteristic dimensions of Ni nanoparticles varying in the range 40 – 60 nm. After 60 min of deposition Ni continuous nanowires of 10 μm length have been formed. XRD analysis confirmed polycrystalline structure of Ni in the PS template with the preferential orientation along <111> axis. Also some amount of Ni2Si silicide was formed, which diffraction peak at 2 $\Theta \approx 33^\circ$ was especially pronounced for the small deposition times. The possible mechanism of nickel silicide formation during the electrochemical process has been discussed. It has been supposed that the presence of amorphous silicon on the pore walls facilitates the diffusion of Ni inside silicon matrix with subsequent nickel silicide formation without heating. The idea has been confirmed by the fact that on crystalline silicon the formation of nickel silicide was not observed. The magnetic properties have been investigated by studying the temperature dependence (77−700 K) of the specific magnetization σ. The measured σ values were lower with respect to that of bulk Ni. The effect has been explained by the influence of uncontrolled formation of nickel silicide, which causes, after heating, larger irreversibility of σ (T) curves for samples with less deposition time. The obtained σ (T) dependencies allowed them to determine the Curie temperature, TC, which for small deposition times of Ni was lower (575 K) with respect to the bulk Ni (630 K). This is caused by the influence of dimensional effects on TC value. The performed ferromagnetic resonance measurements for arrays of Ni NWs in PS templates with different porosities P allowed obtaining the crossover of magnetic anisotropy type from "easy axis" for P< 0.7 to "easy plane" for P>0.7. This crossover is induced by the growth of dipole-dipole interaction between closely packed Ni NWs with P growth. The magnetization M versus magnetic field H measurements performed for samples with P = 0.7 revealed the dominant role of the shape anisotropy of Ni NWs. It was confirmed both by the analysis of the squareness of the M(H) hysteresis loops and anisotropy field obtained within the random anisotropy model for the magnetic field oriented either parallel or perpendicular to the NW axis.

In Chapter 4, the finite-difference time-domain (FDTD) method is used to study the long-wave phonon scattering at geometric imperfections of two-dimensional nanowires. The transmission/reflection spectra of the phonons of different polarizations (bending and compression phonons) are calculated for two kinds of defects: retaining and not retaining the waveguide symmetry. These calculations are relevant to the description of ballistic phonon thermal transport in mesoscopic systems at low temperatures. The elastic energy propagation in a nanowaveguide between two massive blocks is studied by FDTD method. The energy field realized in the system in the process of formation and propagation of the energy flow is calculated. It is shown that in mesoscopic systems the field configuration is determined by the sample geometry and the boundary conditions. Based on the calculations and analysis performed, the conclusion is drawn that in the steady-state regime of energy propagation between two massive blocks through a small-section waveguide with a free surface, energy propagates in a narrow surface layer.

In: Nanowires
Editor: Luke J. Wilson, pp. 1-57
ISBN: 978-1-63117-855-9

Chapter 1

ZnO NANOWIRES: SYNTHESIS APPROACHES AND ELECTRICAL PROPERTIES

Carminna Ottone[1,2], Marco Laurenti[1,2], Paolo Motto[3], Stefano Stassi[2], Danilo Demarchi[1,3] and Valentina Cauda[1,*]

[1]Center for Space Human Robotics,
Istituto Italiano di Tecnologia, Torino, Italy
[2]Department of Applied Science and Technology, Torino, Italy
[3]Department of Electronics, Torino, Italy

Abstract

In the present chapter we aim to overview the synthesis techniques and growth mechanism of zinc oxide (ZnO) nanowires and their related properties. In particular, here we review both wet and dry synthesis techniques reported in the literature, such as hydrothermal and sol-gel routes, template-assisted growth, electrodeposition process, chemical vapor deposition (CVD), pulsed laser deposition and RF magnetron sputtering.

According to each synthesis technique, the chapter reports not only the detailed procedure, but also the advantages and disadvantages, analyzing in particular the properties of ZnO nanowires related to their microstructural and morphological features. The functional properties of ZnO nanostructures are also overviewed with a particular attention to their electrical properties and the related

* E-mail address: valentina.cauda@iit.it. Tel. +39 011 0903436, Fax. +39 011 0903401. (Corresponding Author)

potential applications, exploiting either the semiconductive and piezoelectric behavior of ZnO nanowires. A comprehensive review is here focused on their use as field effect transistor, energy nanogenerators, pH and gas sensors, as well as the use of ZnO nanowires for energy conversion, like photovoltaic cells and water splitting photoelectrochemical cells.

1. Introduction

Zinc Oxide (ZnO) is a metal oxide semiconductor of group II-IV whose nature resides at the borderline between covalent and ionic semiconductors [1]. In the last decade it has raised several attention in the research field since it possesses promising catalytic, electrical, electronic and optical properties [1]. It can be easily prepared in different shapes and sizes at both the micrometric and nanometric scale, including the forms of micro- and nanowires, nanobelts, nanotubes, micro- and nanoparticles, multipods, tetrapods, and flower-like microstructures. This different variety of morphologies has thus attracted considerable attention for potential application in solar cells, nanogenerators, field effect transistors, gas sensors and other electronic micro- and nanodevices [2-4].

ZnO can exist in three different crystal structures, such as rock salt (B1), zinc blende (B3) and wurtzite (B4). B1, B3, and B4 denote the structure report designations for the three phases, as reported in Figure 1.

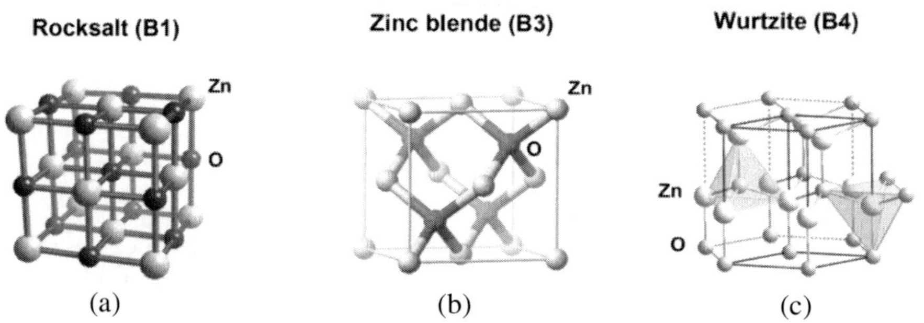

Figure 1. The crystal structures of ZnO: (a) rocksalt, (b) zinc blend and (b) wurtzite unit cells. Grey spheres correspond to the zinc atom, black spheres to the oxygen one.

The wurtzite structure is the most thermodynamically stable phase under ambient conditions. In contrast the cubic rock salt (or Rochelle salt) symmetry is synthesized at relatively high pressures. The zinc blende structure could be stabilized just by growing it on cubic structures. ZnO is otherwise called Zincite in the mineral form, showing a tetrahedral wurtzite-like crystal structure of ZnO_4, however it is extremely rare in nature.

The wurtzite structure has an hexagonal symmetry, constituted by tetrahedral units where each zinc atom is at the center and four oxygen atoms at the edges (see Figure 1c). All tehrahedra are oriented in one direction and their stacking produces the ZnO hexagonal symmetry. Overall, the wurtzite is constituted by alternating planes of oxygen and zinc atoms and do not display any center of inversion. Therefore wurtzite is intrinsically asymmetric along the *c*-axis, or the [001] direction, thus being an anisotropic crystal. The most stable crystal habit is therefore a hexagonal prism elongated along the *c*-axis (Figure 2). Because of the alternating oxygen and zinc planes in the wurtzite structure, the hexagonal prism always have at the one end a (0001) oxygen plane and a zinc plane at the other end. This implies either a negative and a positive polarities at the two ends, i.e. the ±(0001) planes, of the ZnO crystal.

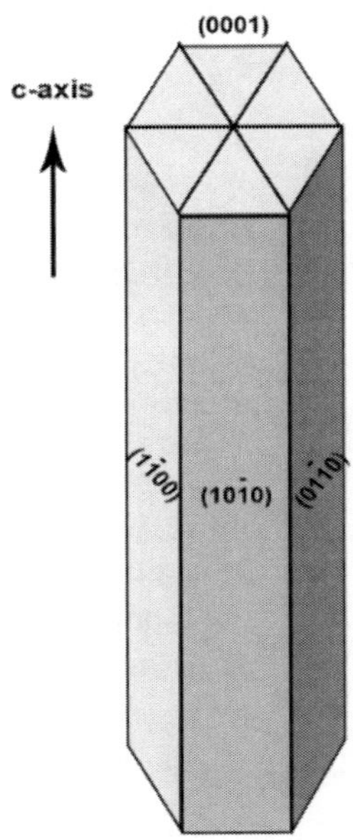

Figure 2. Wurtzite hexagonal crystal structure of ZnO.

The ±(0001) surfaces show therefore the highest surface energy and are metastable. In contrast, the two lateral faces of the hexagonal prism are "low-symmetry" non-polar faces, being the most stable ones, since they show 3-fold coordinated atoms. It is therefore clear why the highest growth rate of ZnO nanocrystals takes place on the (001) face, thus along the *c*-axis direction, leading to anisotropic crystal structures.

The intrinsic polarity of ZnO, in particular in the form of micro- and nanowires, nanobelts or nanotubes leads to piezoelectric properties [5]. In addition, ZnO is intrinsically a n-type semiconductor material showing high electron mobility, high thermal conductivity, wide bandgap (about 3.3 eV), and large exciton-binding energy (~60 meV) [4, 6-7]. It adsorbs the light in the UV range of 200–350 nm and emits in the near UV and visible range from 500 to 600 nm. Moreover, it shows a superior fatigue resistance, high isoelectric point (9.5), charge transfer properties, biocompatibility and low impact on the environment at the end of life cycle [2, 8].

2. ZnO Nanowires Synthesis

ZnO is a very well-known material and several synthetic methods have been reported up to now in order to grow ZnO nanostructures with different morphologies. Here we will focus only on the synthesis of ZnO nanowires, which can be in general divided into either wet-chemical and dry techniques. Among the first category, the solution-based processes consist in the sol-gel methods, such as the aqueous chemical growth [9-10] and oriented crystallization [11], the electrodeposition technique [12-13] and the template-assisted growth [14]. All these methods allow to grow ZnO nanowires at mild reaction temperatures and low costs, also with the possibility of scaling up the product yield.

Among the dry-synthesis techniques, the most used methods in the literature are the vapor-phase transport processes [15-17], such as chemical vapor deposition (CVD) [18], metal-organic chemical vapor deposition (MOCVD) [19], whereas other important methods include pulsed laser deposition (PLD) [20], and Radio Frequency (RF) magnetron sputtering [21]. The great advantages of these dry techniques consist in the high quality and level of crystallinity of the final ZnO nanostructure. In addition, high purity, absence of contaminants, and in some cases catalysts-free growth of ZnO nanowires can be performed.

In the following, these techniques are discussed in details.

2.1. Wet-Chemistry Processes

ZnO can be easily prepared by solution-phase synthesis, being a low cost and low temperature method, resulting in high volume production of ZnO nanostructures, and showing the possibility of selecting different final morphologies by varying the reactant concentration and nature.

In general ZnO is formed from the hydrolysis and condensation of a zinc salt in a basic solution. Commonly, zinc precursors like the inorganic compound zinc nitrate Zn(NO3)2 and the organic one zinc acetate Zn(OOCCH3)2 (also written as Zn(OAc)2) are employed. The bases used are generally NaOH, KOH, ammonium hydroxide NH4OH or hexamethylenetetramine (CH2)6N4. During the reaction, the zinc precursor hydrolyzes into Zn2+ ions which reacts with the OH- anions of the base (Eq. 1). Different hydroxyl species can be formed [22], including ZnOH+(aq), Zn(OH)2(aq), Zn(OH)2(s), Zn(OH)3-(aq) and Zn(OH)42-. Then, the hydroxide intermediate condenses into the oxide form (Eq. 2).

$$Zn^{2+} + 4OH^{-} \leftrightarrow [Zn(OH)_4]^{2-} \qquad \text{(Eq. 1)}$$

$$[Zn(OH)_4]^{2-} \leftrightarrow ZnO + H_2O + 2OH^{-} \qquad \text{(Eq. 2)}$$

The overall reaction can thus be written as:

$$Zn^{2+} + 2OH^{-} \leftrightarrow ZnO + H_2O \qquad \text{(Eq. 3)}$$

These chemical reactions in aqueous systems are usually equilibrium reactions, which tend to evolve in the direction of the free energy minimization of the entire system. As mentioned before, the highest energy surfaces are the ± (0001) polar planes, terminating with either zinc (Zn^{2+}) or oxygen (O^{2-}) atoms. Therefore the addition of the new molecules from the precursor hydrolysis happens on these polar surfaces in order to minimize the surface energy. In particular, the adsorption of molecules happens by the formation of a monolayer, having the opposite polarity of the pre-existing (0001) plane. In this way a Zn^{2+}-terminated surface transforms into a O^{2-}-terminated one, or viceversa. Therefore the growth of ZnO micro- and nano-structures happens along the *c*-axis direction, leading to anisotropic structures.

In general the wet-chemistry processes have the advantage of using low temperature and reagents with low cost with respect to the vapor-phase synthesis, employing high temperatures and expensive instrumentations. Wet syntheses show high flexibility, giving a broad control on the chemical components and final composition, final nanostructure morphologies, large area coating of the substrate, if any, or large scale production of ZnO in powder form. In addition, when the chemical reaction are carried out in water as solvent, the synthesis is quite eco-friendly and does not requires high power sources or high vacuum. In this kind of synthesis, generally no further thermal treatment is required to obtain crystalline wurtzitic ZnO. However, the wet-chemistry processes show some disadvantages, such as the risk of reactant and solution contaminations, reducing the crystallographic quality of the ZnO sample, and relatively long growth time to obtain micrometer-long nanowire structures.

2.1.1. Hydrothermal Growth

From the growth mechanism explained above, different ZnO micro and nano-wire structures can be obtained by simple hydrolysis and condensation method in basic solutions [23-24].

For example micro-sized wires ranging between 10-15 μm in length and about 300-500 nm in diameter can be obtained with an ease and highly reproducible hydrothermal synthesis at low temperature, about 70 °C (Figure 3). The reaction involves the hydrolysis of zinc nitrate hexahydrate $Zn(NO)_3 \cdot 6H_2O$ with potassium hydroxide (KOH) in a specific molar ratio for 4 hours in water solutions. The obtained product is a white powder which need to be thoroughly washed with water by filtration or centrifugation in order to remove the unreacted chemicals, if any, and neutralize the basic pH of the growth solution. The obtained wires show an hexagonal cross section and flat tips. X-ray diffraction pattern carried out on the obtained powder shows sharp and intense reflections, confirming the high level of crystallinity of the product. The reflections are attributed to the wurtzite crystalline phase.

Due to the preferential growth along the c-axis, single crystalline wurtzitic nanowires from few nanometers up to hundreds of nm can be obtained with an approach called oriented attachment [11]. Here crystalline nuclei of ZnO are prepared in organic and dehydrated solvent medium starting from Zn(OAc)2 as zinc precursor and a strong base, either KOH or NaOH. The nuclei grow into small nanoparticles of few nanometers and then coalesce attaching one to each other at the (0001) crystallographic plane, which is the most reactive surface, as mentioned above. This oriented attachment of single crystalline nanoparticles leads to partially fused dimers (Figure 4a), then to oligomers, and finally to single crystalline nanorods of hundreds of nanometers in length (Figure 4b).

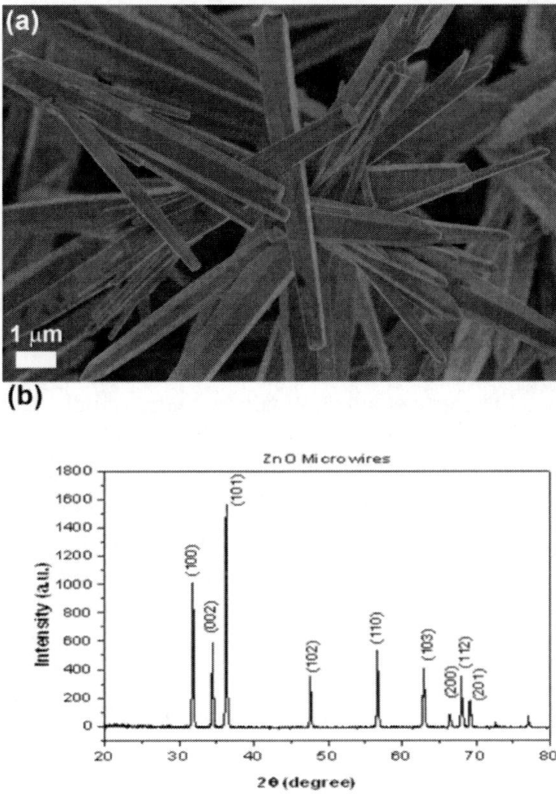

Figure 3. (a) Field Emission Scanning Electron Microscopy (FESEM) image and (b) X-Ray Diffraction (XRD) pattern of ZnO microwires, synthesized by low temperature hydrothermal method.

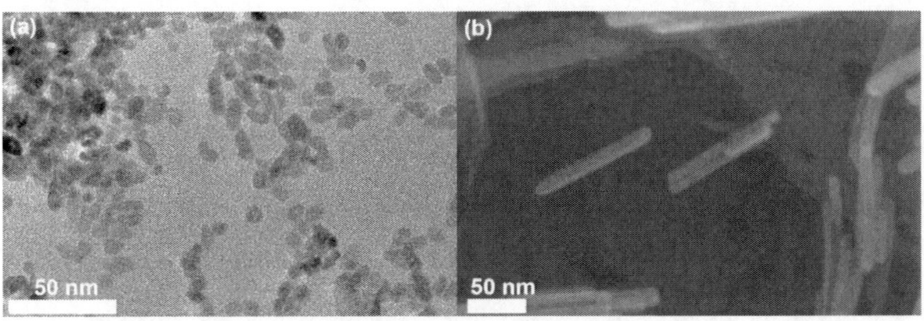

Figure 4. High Resolution Transmission Electron Micrograph (HR-TEM) of (a) ZnO nanoparticles and dimmers; (b) nanowires prepared by the oriented attachment approach.

2.1.2. Aqueous Chemical Growth of Vertically Oriented Nanowires

ZnO nanowires with different diameters and lengths can be also easily obtained as an array of vertical nanostructures grown on a substrate. This oriented deposition takes place by an hydrothermal aqueous chemical process, in which the substrate is immersed in a growth bath consisting of a zinc precursor and hexamethylenetetramine (HMT), NaOH or NH_4OH as source of OH^- anions [8].

The most used substrates are generally constituted by silicon wafers, possibly covered by a thin film of conductive metal, GaN epilayer grown on sapphire, glass slides or Transparent Conductive Oxide (TCO) layers on glass substrates, such as ITO (indium tin oxide) or FTO (fluorine-doped tin oxide).

The substrate is then immersed vertically or horizontally, face down to the bottom of the reactor, in the growth bath and the closed vessel is placed either under magnetic stirring or in static conditions at temperatures varying between 70 and 90 °C for several hours (generally ranging from 1 h to 24 hours).

The overall chemical reaction taking place during the ZnO nanowires growth is that reported by Equation 3, where the hydroxyl anions are supplied by the base used. In the case of HMT, a nonionic cyclic tertiary amine, it decomposes into formaldehyde and ammonia and acts as a pH buffer, as follows:

$$(CH_2)_6N_4 + 6H_2O \leftrightarrow 4NH_3 + 6HCHO \qquad \text{(Eq. 4)}$$

$$NH_3 + H_2O \leftrightarrow NH_3 \cdot H_2O \leftrightarrow NH_4^+ + OH^- \qquad \text{(Eq. 5)}$$

The aqueous chemical growth has many advantages since it is low cost, the process is carried out at low temperature and can be scalable, thus guaranteeing the coverage of broad surface areas (up to some cm^2) by the ZnO nanowire array. In addition, it does not require high pressure containers and is also entirely recyclable, safe and environment friendly since water is used as solvent [10]. In this synthesis no additional heat or chemical treatments are required to achieve the wurtzite crystalline structure of ZnO. In particular single-crystalline one-dimensional (1D) ZnO nanowires arranged in a vertical array are obtained throughout the whole surface of the substrate exposed to the growth solution (Figure 5). The length of the nanowires can be opportunely tuned by adjusting the reaction time of the substrate in the growth bath. Therefore longer the time, longer the obtained nanowires up to the complete depletion of the reactants. In this case, when the zinc source is completely exhausted in the reaction mixture a series of side reactions takes place, leading to the erosion of the formed nanowires by the excess of OH^- groups. This leads to nanowires with concave or pencil-like tips, as well as to the Ostwald ripening enlarging the diameter of the formed nanowires [25].

To avoid these side effects several authors have proposed the growth of nanowires for longer times by refreshing at selected time intervals the growth solution with a new one [9, 26]. In this way nanowires up to 30 μm in length and with a high aspect ratio (i.e. length to diameter ratio) were obtained [9]. This process is necessary for certain applications where a broad surface area of the ZnO nanostructures, and thus a high length, are required, i.e. semiconducting applications in dye-sensitized solar cells (DSSCs) or enhanced piezoelectric properties [26].

In particular highly crystalline, long and well vertically aligned ZnO nanowires can be synthesized on crystalline ZnO thin films deposited on the selected substrate. This polycrystalline ZnO film is called seed layer and is formed by several crystals oriented along the *c*-axis direction. Its role is to drive the nucleation of the ZnO nanowires in the growth bath along the *c*-direction, leading to a controlled and oriented synthesis of vertical ZnO nanowires, uniform in size [27].

The properties of the zinc oxide seed layer can be tuned by adjusting the deposition conditions, the dry or wet process selected to deposit it, the temperature, humidity and pressure [28]. All these factors would deeply influence the final crystalline aspect and orientation of the ZnO seeds, thus affecting the further nanowires growth.

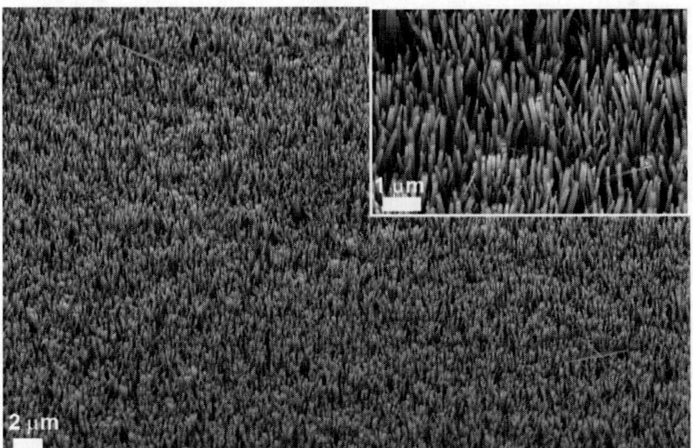

Figure 5. Vertical array of ZnO nanowires on a broad surface area.

The most common techniques used for depositing the seed layer is the sol-gel method, using a precursor solution of zinc acetate in a volatile solvent (either ethanol or isopropanol) which can be deposited on the substrates by either spin-coating, dip-coating or drop-casting techniques [25]. Here the fundamental parameters to be tuned are the solution concentration, the speed and time of the deposition, and the thermal treatment used at the end of the process, which is necessary to crystallize the zinc precursor in a ZnO polycrystalline thin film. Another common process used for the seed layer deposition is the radio-frequency magnetron sputtering technique [29], see also below. In this case, parameters like the sputtering time, the gas pressure, the target used for the deposition are essential to optimize the crystalline quality of the ZnO thin films, which does not require in this case any further thermal treatment.

After the seed layer deposition, the substrate is then immersed in the nanowire growth bath, as explained above. The major advantage of using ZnO seedlayer is that the ZnO nanowires can be grown on whatever substrates, from hard Si wafers (flat, etched, or patterned) or metal layers to flexible substrates, such as polydimethylsiloxane (PDMS), thermoplastic polyurethanes (TPU), paper, organic and carbon fibers and flexible polyimide (Kapton) [8].

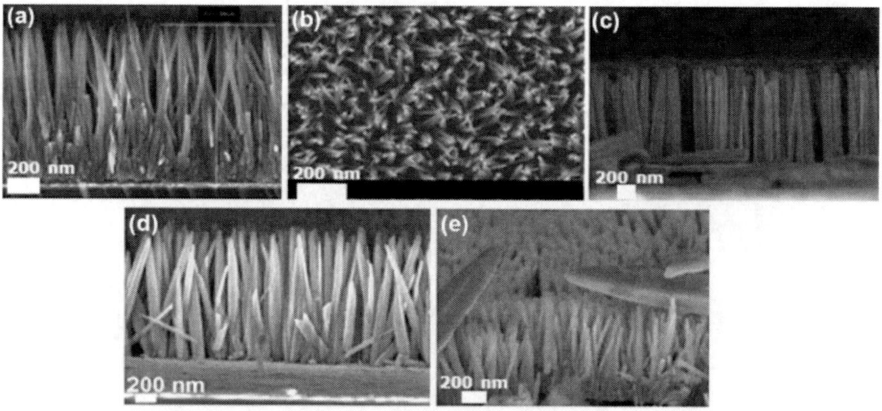

Figure 6. Vertical array of ZnO nanowires grown on seeded substrates. (a) Cross section and (b) top view of nanowires on seedlayer deposited by spin coating on silicon wafer covered with Ti/Au conductive layer; (c) Nanowires cross-sectional view grown on sputtered seed layer on Si/Ti/Au wafer; (d) ZnO NWs from seed layer deposited by dip-coating on FTO/glass substrate and (e) on dip-coated seed layer on flexible polyimide substrate having a thin copper conductive layer.

In general, the growth bath shows a high degree of supersaturation with respect to ZnO, in order to induce the growth on the substrate, either seeded or not. However this implies also the homogeneous nucleation of ZnO particles in the bath, which is not desirable. The homogeneous nucleation contributes indeed to the reactant depletion and the formed crystals can deposit on the as-grown nanowires contaminating them and reducing the crystallographic quality of the final sample. In order to reduce the homogeneous nucleation and further improve the growth rate and the anisotropy of the 1D nanowires, recently the addition of capping agents as polyethylenimine (PEI), together with ammonium hydroxide, to the Zn^{2+} supersaturated growth solution was proposed by Xu et al. [9]. NH_4OH is able to form coordination complexes with the zinc cations Zn^{2+}, which remains therefore in solution without reacting with the OH^- anions. In this way the homogeneous nucleation is suppressed.

PEI is a nonpolar low molecular weight polymer, possessing a large amount of amine groups ($-NH_2$) which can be protonated over a broad pH range, from 3 to 11. It can therefore adsorb to the negatively charged lateral facets of ZnO clusters due to electrostatic interactions, thus preventing the lateral growth of the ZnO nanowires and promoting only the anisotropic growth along the *c*-axis. In addition, the PEI has also a similar coordination effect on Zn^{2+} cations as ammonium hydroxide, thus further suppressing the homogeneous nucleation. In these conditions, several works reported on the growth of ZnO wires with high aspect ratio on seeded substrate at a reasonably high growth rate (approximately 1 μm per hour) [25, 29-30]. Anyway, the growth of longer nanowires requires the refreshing of the growth solution, as mentioned above.

2.1.3. Electrodeposition

The formation of nano- and micro-wire arrays of ZnO on conductive substrates can be obtained from the electrochemical deposition or the so called electrodeposition. During this process, an oriented diffusion of charged reactive species through the solution takes place under the application of an external electric field. The electrodeposition is a versatile, low cost, easy and scalable method. It also works at low temperature (<100°C) and the diameter and length of the nanowires can be easily tuned by varying the process conditions, such as the voltage, time, charge density and precursor concentration. Nevertheless, the limitation of this process is the requirement of a conductive substrate to work as electrode in the electrochemical cell. However, a thin layer of conductive material, such as gold or platinum, can be generally deposited on non-conductive substrates in order to render them conductive and ensure the electrical continuity.

Gold-coated silicon wafers and FTO or ITO glasses are commonly used as substrates.

To carry out the electrodeposition process, a standard three-electrode setup is typically used. A conductive substrate acts as the working electrode, a Platinum wire or sheet serves as the counter electrode and a saturated calomel (SCE) or Ag/AgCl electrode as the reference electrode. The potential values are recorded with respect to the reference electrode. A constant voltage source is maintained in order to guarantee a constant driving force.

The ZnO electrodeposition consists on the reduction of an oxygen precursor, such as dissolved molecular oxygen, nitrate ions (Eq. 6) or hydrogen peroxide (Eq. 7). The cathodic electrodeposition of ZnO nanowires has been reported from different solutions of zinc salts, such as zinc chloride ($ZnCl_2$) and zinc nitrate ($Zn(NO_3)_2$) [31-34]. The electrochemical reduction of O_2 occurs by a two (Eq. 8) or a four (Eq. 9) electron process, as a function of the electrolyte and cathode properties. In general, the voltage range where the ZnO reduction occurs varies between 0.5 V and 2 V, depending on the electrolyte and geometrical cell dimensions. The different oxygen precursor reduction processes during the electrodeposition of ZnO are supposed as follows.

Nitrate: $$NO_3^- + H_2O + 2e^- \rightarrow NO_2^- + 2OH^- \quad \text{(Eq. 6)}$$

Hydrogen peroxide: $$H_2O_2 + 2e- \rightarrow 2OH^- \quad \text{(Eq. 7)}$$

Molecular oxygen: $$O_2 + 2H_2O + 2e^- \rightarrow H_2O_2 + 2OH^- \quad \text{(Eq. 8)}$$

$$O_2 + 2H_2O + 4e^- \rightarrow 4OH^- \quad \text{(Eq. 9)}$$

The electrochemical reduction of the oxygen precursors into hydroxide ions induces an increase of the local pH close to the cathode. As similarly reported for the synthesis of ZnO nanowires through the wet-chemistry processes, Zn^{2+} and OH^- ions react together leading to the precipitation of ZnO and the deposition on the surface of the cathode, as summarized from Eq. 1 to Eq. 3. In addition, directing agents such as hexamethylenetetramine (HMT), ethylenediamine (EDA) or NH_4F are also added to the electrolyte solution in order to obtain nanowires instead of a continuous ZnO film [35-36].

The growth of ZnO nanowires starts with the formation of nucleation sites adherent to the substrate. The charge density increases among the ZnO growth. At lower stages of the ZnO formation, the charge density is lower than 2 C/cm^2. With this value there is no preferential growth in the longitudinal direction, therefore pyramidal structures are first formed. Then, when the charge density increases, wider pyramidal crystals are formed, which subsequently results in thicker nanowires. At further stages, the incoming Zn^{2+} species find first the tip of the nanowires, therefore the most straight and vertical nanowire structures consume the arriving Zn^{2+} ions, whereas the tilted ones have less probability of reacting with the cations. The Zn^{2+} ions react with the OH^- ions adsorbed on the top faces of the growing nanowires. However, the Zn^{2+} ions do not reach the tips of the most tilted nanowires, quenching their growth. At the end of the process, among the irregular nanowire lengths, the straight and vertically oriented nanowires result to be the longest ones.

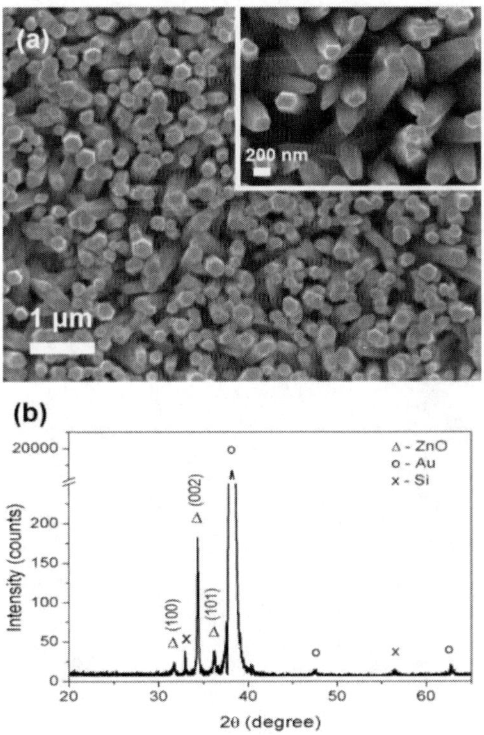

Figure 7. (a) Top view FESEM image and (b) XRD pattern of ZnO nanowires synthesized through the electrodeposition approach and grown on Au-coated silicon wafer. The inset shows a higher magnification image.

In the specific case of using molecular O_2 as OH^- precursor, it was demonstrated that the principal parameter affecting the longitudinal growth is the balance between the diffusion of the Zn^{2+} and O_2 species. The electrochemical reduction of O_2 appears to be the most important process influencing the growth of the ZnO nanowires. In particular, when the diffusion of the Zn^{2+} species is inferior to the diffusion of O_2, the wire configuration is favored.

2.1.4. Template-assisted Approach

Template-assisted approach is a versatile method for synthesizing nanowires or nanorods of various materials. It is readily scalable to mass production. Nanoporous materials with small cylindrical pores act as templates for obtaining one dimensional (1D) nanostructures. The most common templates are anodic aluminum oxide (AAO) membranes [37-38] (with pore diameter ranging from 20 to 500 nm) and track-etched polycarbonate membrane. Among the less common porous hosts, mesoporous silica [39] (i.e., micelle-templated porous SiO_2 with pore sizes of 2–15 nm), porous silicon, and carbon nanotubes [40] are used. The current technology permits to obtain templates with a narrow pore size distribution [41]. Thus, through the template-assisted approach, nanowire arrays with well-known geometrical parameters are fabricated. The diameter, density and length of nanorods and nanowires are those of the porous host and thus easily controllable independently.

Figure 8. ZnO nanowires obtained upon (a) electrodeposition and (b) wet impregnation into porous alumina matrices. The images show the alumina membrane filled by the ZnO material after template dissolution.

Template-based growth requires the use of precursor solutions, or sols to fill the pores by simple wet impregnation or by electrodeposition, as previously explained in this chapter [33, 37, 42]. ZnO nucleation occurs in the solution diffusing into the template channels. Then, the ZnO crystals grow adopting the pore morphology and shape. When combining the template-assisted approach with a wet chemical method, the growth can also proceed from the channel walls inwards to the channel center, thus having the possible disadvantage of pore blocking. Therefore the major drawback of the template-assisted approach is the difficulty to ensure the complete filling of the pores [43].

When ZnO is synthesized by combining the electrodeposition and template-assisted method, the host must be constituted by an electrical insulating material. In this case, a thin layer of a conductive material is deposited on one side of the template membrane in order to favor the nucleation of the crystals at one side, i.e. the bottom, of the template channels. The growth proceed then from the bottom to the top of the channel, thus avoiding pore blocking [43]. The structural and surface properties of the formed ZnO nanowires strongly depend on the quality of the host pore walls. The presence of defects, asperities, irregularities at the pore walls may result in polycrystalline ZnO nanowires with a rough and irregular surface [32]. The template must fulfill the requirement of having compatibility with the used solution, i.e. chemical and thermal inertness. Depending on the final application, the nanowire array can be used together with the template or can be separated by performing a selective dissolution of the template.

2.2. Dry-Methods

As reported above, ZnO NWs can be prepared by different dry-synthesis methods. Depending on the process through which the desired material is formed, dry methods can be classified as physical vapor deposition (PVD) or chemical vapor deposition (CVD) processes. PVD and CVD methods are based on the formation of a vapor phase and the consequent condensation of the species to be deposited onto the substrate. PVD techniques involve the extraction of atoms and/or molecules starting from a solid source and the vapor phase is formed through physical processes like evaporation or sputtering. On the contrary, in the case of CVD techniques, a chemical reaction occurs between solid or gaseous precursors, and is exploited for the formation of the vapor phase. Vapors can be carried out either in vacuum or using inert gases. The formation process of the material starts with the adsorption of atoms and/or molecules on a surface (adatoms) through physical or chemical interactions. In the first case (physisorption), electrostatic interactions like Van der Waals forces are responsible for the adsorption of the atoms onto the surface. In the second case (chemisorption), the adsorption takes place making chemical bonds between the atoms and the surface. After the adsorption of the atoms, the nucleation process can take place, during which different adatoms can aggregates to each other, giving rise to the formation of islands. Depending on the way these islands aggregate to each other, different growth modes can be identified: the island growth mode, the Frank–Van der Merwe mode, and the Stranski–Krastanov mode. In the first case, the interaction between islands is predominant with respect to the one between adatoms and the surface. In the second case, the dominant process is the one with the surface while in the last case the two interactions (adatom/adatom and adatom/surface) compensate each other. The final morphology and the micro/nano structure of the resulting materials is influenced and partially determined by the particular growth mode which is favored during the deposition process.

2.2.1. Chemical Vapor Deposition (CVD)

Among the large variety of dry methods for growing ZnO nanostructures, the chemical vapor deposition (CVD) is by far one of the most investigated. The versatility of CVD processes is mainly related to the possibility of growing a large number of compound materials such as oxides, carbides and nitrides [44]. CVD can be considered as a vapor-phase transport technique and the process is based on the deposition of atoms and/or molecules upon heating the substrates, starting from a chemical reaction in the vapor phase. The entire process can be divided into five steps. The first and second steps concern the diffusion of the reactants to the substrate and the adsorption onto the surface. The third one involves the surface chemical reactions, leading to the deposition of the solid material. Gaseous by-products begin a desorption process from the surface and finally start to diffuse into the stream. On the basis of the method used for the activation of the chemical reactions, CVD processes can be properly classified. When the activation energy comes from a pyrolysis process, the growth of the materials is carried out at very high temperatures and the technique is usually called thermally activated CVD and will be properly described in the following. It can be divided in atmospheric pressure CVD (APCVD) or low pressure CVD (LPCVD). In the first case, depositions are carried out at atmospheric pressure while in the second case depositions are performed at pressure levels around 10^{-3} Torr. Other CVD techniques are also present, like the metalorganic CVD (MOCVD), which will be discussed in paragraph 2.2.2, and the laser/ photo CVD, where the chemical reactions are activated through photon irradiation, respectively with the aim of heating the substrate or breaking the chemical bonds of the precursors.

Thermally activated CVD is one of the most diffused ways for producing a great variety of different ZnO nanostructures, such as nanowires, tetrapods, combs, springs, rings, trees, belts and rods [2, 16, 45]. It is a very simple process, where the vapor phase is obtained vaporizing condensed or powder source materials at elevated temperatures. The resultant vapor phases condensate onto the substrate to form the desired products. The processes are usually carried out in a horizontal tube furnace, generally made of quartz or alumina. A gas supply, a control system and a rotary pump (except for APCVD) are also present. An inert gas is used as carrier for the reactive species coming from the source materials, which are normally loaded on an alumina boat positioned inside the tube. On the basis of the source materials, oxygen is added to the inert gas at different concentrations. The morphology of the final material can be controlled by properly varying several processing parameters like temperature, pressure, carrier gas, gas flow rate, substrate and evaporation time. The source temperature is selected taking into account the volatility of the source materials and it is usually slightly lower than its melting point. The type of product that will be obtained is strictly dependent on the drops of the substrate temperature with the distance of its location from the position of the source. Another fundamental variable is the concentration of oxygen, which influences the volatility of the source and the stoichiometry of the vapor phase.

CVD equipments are generally simple and relatively cheap, not requiring the presence of a high-vacuum system, allowing the deposition on wide-area substrates, also with irregular shape, with a high deposition rate and efficiency. However, the absence of a fully clean atmosphere during the growth get the materials quality worse and the high growth temperatures required from this kind of techniques, usually not lower than 500°C, are very restrictive, especially when deposition on polymers are intended to be done. Concerning the thermally activated CVD, depositions performed at atmospheric pressure allow to reach higher deposition rates while materials grown by LPCVD are characterized by a better uniformity and purity, despite the required higher temperatures and lower deposition rates. Figure 9 shows ZnO NWs grown on seeded substrates with APCVD and LPCVD techniques. As it can be seen, when APCVD is performed, ZnO NWs grow with different morphologies and uniformity depending on the distance from the Zn source (Figure 9(a) and 9(b)). This loss in uniformity is not observed in the case of LPCVD as shown in Figure 9(c). In this case, a good coverage and uniformity of the as-grown NWs can be observed over the entire surface area.

ZnO nanostructures can be grown by thermally activated CVD using either ZnO or metallic Zn source, with the addition of oxygen in the gas stream. The use of different ratios and mixtures of Zn and ZnO powders in the presence of oxygen under different growth temperatures changes the morphology and allows the formation of different nanostructures. The growth of ZnO NWs with or without using a metal catalyst has already been reported in the literature [46] and a lot of works show that gold, both in the form of a continuous layer or nanoparticles, is the best candidate as metal catalyst for growing well-aligned, *c*-axis oriented ZnO NWs [47]. With pre-seeding substrates with catalysts, ZnO nanostructures can be grown with a very good control on size and homogeneity [48]. However other works underlined the importance of using a thin ZnO layer acting as crystallization seed for the following growth of NWs, with the possibility of controlling and improving its crystal orientation, size and homogeneity [49] (see also Paragraph 2.1.2).

ZnO seed layers can be prepared with different methods and those obtained by sputtering or sol-gel techniques seem to be the most promising ones [25, 29-30]. Some examples of FESEM images and XRD spectra of ZnO NWs grown by thermally-activated CVD on seeded substrates by the sol-gel method are shown in Figure 10.

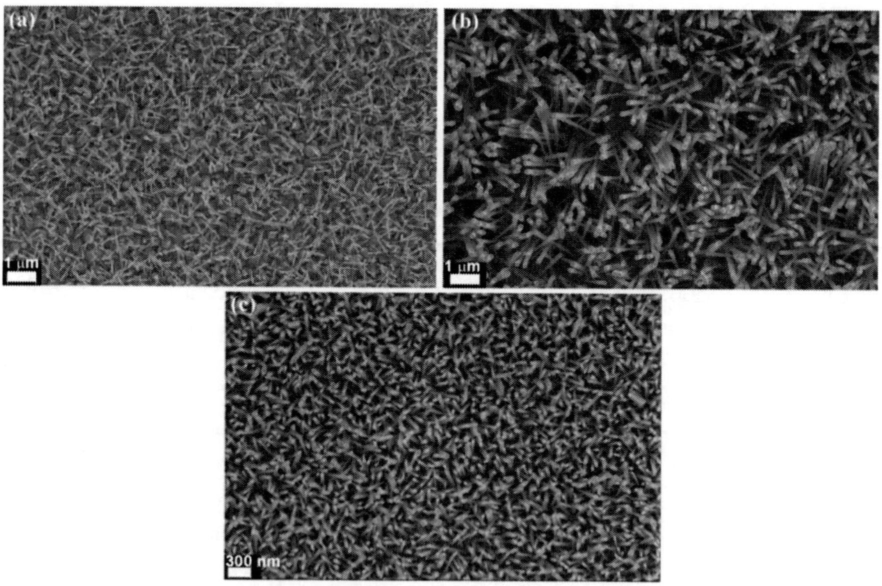

Figure 9. FESEM images showing ZnO NWs grown by thermally-activated APCVD on seeded substrates (a) far from the Zn source and (b) near the Zn source. (c) ZnO NWs grown by thermally-activated LPCVD.

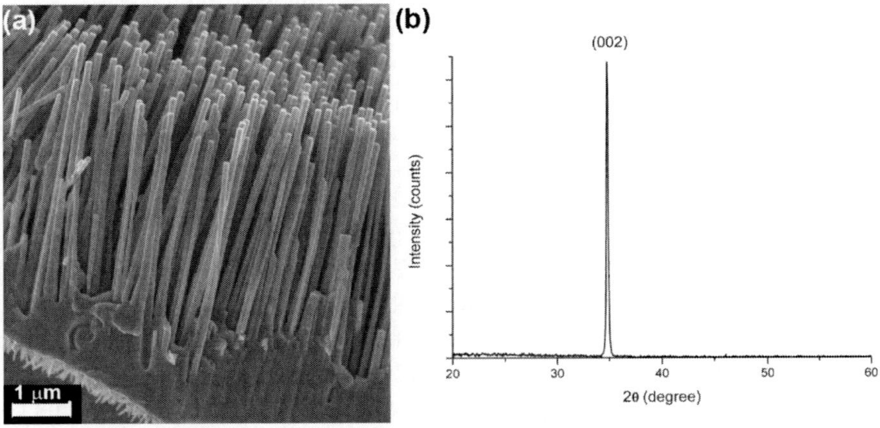

Figure 10. FESEM image (cross-section) and XRD spectrum of ZnO NWs grown by thermally-activated APCVD.

2.2.2. Metal-Organic Chemical Vapor Deposition (MOCVD)

The metal-organic chemical vapor deposition (MOCVD) technique allows the growth of compound materials, exploiting the pyrolysis of various combinations of organometallic compounds and hydrides [44]. The MOCVD process takes place in a cold-wall reactor, and can be divided in four regimes. Reactants are introduced in the chamber and mixed together. During this step, complications about gas phase reactions and reactant diffusion can affect the quality of the as-grown material. Generally the sources are first introduced as vapor phase constituents at approximately room temperature, but thermally decomposed at temperatures generally comprised between 200 and 400°C by a hot susceptor [44]. The most basic parameters that can be varied during a MOCVD process are the susceptor temperature and the input reactant molar flows.

Typically, a MOCVD system is constituted by reactant storages, gas manifold, the reaction chamber and the system for the exhaust management, generally constituted by pumps and scrubbers for the expulsion of exhaust species from the reactor.

MOCVD is a technique characterized by the possibility of growing a large variety of compound semiconductor materials on large area substrates, with a high growth rate and efficiency, and large uniformity [44]. However the use of metalorganic precursors is a quite deleterious disadvantage, because of problems related to their high degree of toxicity and problems related to the disposal of toxic exhausts. Also the necessity of operating at elevated temperatures is a limitation, especially in the possibility of using substrate materials sensitive to high temperatures like polymers.

The MOCVD allows the possibility of producing long and vertically aligned ZnO NWs [50]. It is a bottom-up technique, where the nature of the substrate plays a vital role for the determination of NWs dimension and alignment. Juàrez et al. reported on the MOCVD growth of ZnO NWs starting both from Au sputtered layers and nanoparticles, starting from bis(2,2,6,6-tetramethyl-3,5-heptanedionate)zinc(II), $Zn(TMHD)_2$, as Zn precursor [51-52]. However the possibility of growing single crystal, well-aligned ZnO NWs was also reported in other works without using a metal catalyst [19, 53-54]. Generally diethylzinc (Et_2Zn)or dimetylzinc (Me_2Zn) are the most diffused Zn precursors used for growing ZnO NWs by MOCVD, together with oxygen and/or H_2O. In this case strong reactions occur in oxidizing ambient, resulting in the deposition of particulates upstream the susceptor. This problem can be avoided using other oxidizing elements like carbon dioxide or nitric oxide. However, a lower growth rate is obtained in this case. To overcome the problem of gas-phase powder formation, low partial pressures of high-purity oxygen can be used. There is also a strong influence of the growth temperature on the obtained final ZnO nanostructures. In particular, it is reported that film-like structures are obtained when the temperature is around 360°C. At higher temperatures (500°C) ZnO NWs can be formed. The reason is related to the mobility of precursor species such as [MeZn(OMe)] on the substrate surface. Zn adsorbate must have a sufficient surface lifetime to migrate to the tip of the growing nanowires. The surface lifetime is dependent on the substrate temperature. At low temperatures (300-450°C) Black and co-workers [55] reported that the alkylzinc alkoxide species have not sufficient surface mobility for nanowires growth, and the formation of a polycrystalline film with randomly-oriented ZnO crystallites is reported. At 500°C the precursor complex mobility allows an efficient surface diffusion for growing ZnO NWs.

2.2.3. Pulsed Laser Deposition

Regarding the deposition of ZnO-based materials, the pulsed laser deposition (PLD) is a physical vapor deposition process widely used for growing ZnO in the form of thin films and nanostructures [56]. The physical principle of a PLD process is based on the ablation of a solid source (target) when interacting with a laser radiation. The ablated species are carried from the target to the substrate as vapors. The formation of a continuous film on the substrates can take place through the condensation of the vapors on the surface. After that, nucleation and coalescence can start, leading to the formation of a continuous film. Basically, a PLD set-up is made of a vacuum system for reaching suitable high vacuum conditions for depositing high quality thin films. Targets and substrates are positioned such that the plasma plume originated from the target can reach the centre of the substrate surface. Target motion is usually performed during PLD processes, in order to limit particle emission during the ablation of the target. Depending on the kind of materials to be deposited, gases are selected. For example, oxide deposition is usually carried out employing a mixture of argon and oxygen, while nitrides can be deposited in a nitrogen or NH_3 atmosphere.

The PLD technique is a very promising technique for the deposition of high quality materials thanks to the possibility of simply transferring the stoichiometry of the target into the grown structures. However the particulate emission from the target during the ablation process is one of the main issue regarding this deposition method.

Concerning the deposition of ZnO based materials, PLD is widely used for the formation of thin films [57-58]. However some works report about the synthesis of ZnO nanostructures, such as doped and undoped nanowires and nanorods, with or without using a catalyst [59-61]. Generally PLD depositions of ZnO based materials are carried out in a argon or oxygen background atmosphere, using a ZnO target made of pressed and sintered ZnO powders as the ZnO solid source while KrF or ArF excimer lasers are employed as the laser radiation. Hartanto et al. [62] reported that a crucial point for growing ZnO nanorods is the formation of *c*-axis oriented ZnO nanoparticles in the initial stage of the growth, when the typical island-growth mode takes place. During this phase, the horizontal growth is favored, and the surface of the substrate starts to be covered by ZnO nanoparticles. When all the surface is covered, the horizontal growth is saturated and the vertical growth of ZnO nanorods is then observed. The authors also pointed out a dependence of the growth mode from the gas pressure used during the PLD process. In particular they stated that a high oxygen gas pressure, more than 1 Torr, is necessary for the formation of nanoparticles by the condensation of the ablated species during the initial stage, followed by the formation of nanorods. Yang et al. [63] also reported the growth of ZnO NWs by the PLD technique with or without using a catalyst, on metals and semiconducting substrates like copper, tungsten and silicon. Also in this case, the formation of a thin ZnO layer in the early part of the deposition process is believed to play a fundamental role for the subsequent growth of ZnO NWs. There is also a dependence of the kind of nanostructures from the metal used as substrates. It was also reported on the growth of ZnO nanostructures comparing different kind of metal substrates such as copper (Cu) and tungsten (W) [63]. They noticed that

NWs grew only on the W layer, while ZnO microflowers were present on the Cu substrate. This difference in the morphology of the final ZnO nanostructures was ascribed to the formation of large strain from the misfit in lattices between the compared metal substrates and the thin ZnO layer, which leads to structural transitions and different morphologies of the as-grown ZnO nanostructures.

2.2.4. Sputtering

Sputtering belongs to the class of plasma-assisted physical vapor deposition processes, in which a plasma is used as the source of energy for the formation of a vapor phase. Literally, sputtering process is the extraction of atoms from a solid source (target). A noble gas (usually argon) is injected into the deposition chamber and ionized by the application of a proper potential between the cathode (where the target is placed), and the anode (where the substrate holder is placed). The impinging of the ions on the surface of the target allows the extraction of atoms. Once the vapor is formed, condensation on the surface of the substrate can take place and, by nucleation and coalescence, a continuous layer can be formed. The mechanism of sputtering can locally create non-neutral electrical charge regions on the surface of the target, highly reducing the sputtering efficiency. For this reason, depending on the nature of the target material, both direct current (DC) or radio-frequency (RF) signal voltage are used. When conducting materials such as metals are involved, both types of signal can be used. However, when non-conducting materials like oxides, nitrides or carbides have to be deposited, the use of a RF signal is required, in order to neutralize the local charge created during the sputtering process, enhancing the sputtering efficiency. Depending on the nature of the target material and the material to be deposited, sputter deposition can be performed in an inert atmosphere or in a mixture of inert and reactive gases (oxygen). When a reactive gas is used, the process is referred to as the reactive sputtering technique.

Sputter deposition of materials is a high yield production technology, compatible with integrated circuit (IC) processing. The possibility of depositing homogeneous materials on large-area substrates with a high deposition rate, also at low temperatures, makes this technique suitable and advantageous for the deposition on a large variety of substrates, such as metal substrates, silicon, and flexible polymers [64-65]. Sputtering requires the presence of quite costly high-vacuum systems in order to allow the formation of plasma and the presence of a suitable atmosphere, free from any contaminants or pollutants.

Sputtering is conventionally used for the deposition of ZnO nanostructured thin films, and a lot of works described their excellent properties and applications

[66-68]. Generally, ZnO thin films can be deposited starting from both a metallic Zn or a ceramic ZnO target. When starting from a metallic target, a mixture of argon and oxygen has to be injected into the vacuum chamber in order to form and deposit the oxide species. The sputter growth of ZnO nanostructures, such as nanowires, nanorods, nanobelts and nanocables were also reported in literature [21, 69-71]. In particular, Chiou et al. [21] reported about the growth of single crystal ZnO nanowires using sputter deposition. In this case, the presence of a thin copper layer seems to be somehow necessary for growing NWs instead of dense thin films. The influence of the copper layer on the growth of nanowires is thought to be related to the presence of stress concentrators, which favors the occurrence of a Stranski-Krastanov growth mode, leading to the formation of ZnO island-shape particles subsequently used as the seed for the growth of ZnO NWs. When these stress concentrators come to be all occupied by the presence of one-dimensional nanostructures, the Frank-Van der Merve growth mode is the predominant one and ZnO starts to grow in a heteromode, leading to the formation of two-dimensional thin films. Chen et al. studied in detail the relationship between different kind of copper layers and the subsequent growth of nanorods [71]. In particular, copper layers deposited by thermal evaporation, DC sputtering and electroless plating were compared and the influence on growing one-dimensional nanostructures was analyzed. They found that NRs grew only in the case of the electroless plated Cu layer, whereas in the other two cases only ZnO thin films were obtained. They also found a relationship between the growth of NRs and the average thickness and roughness of the electroless plated Cu layer. In particular they stated that there should exist a critical value of these two parameters for which the growth of NRs occurs. Chiou et al. also noticed that the presence and the nature of an interlayer between the silicon substrate and the copper layer may influence the growth of nanowires [21]. Titanium (Ti) and tantalum nitride (TaN) were compared, finding that single crystal nanowires grew only when Ti was used as interlayer while, in the case of TaN, ZnO was present only in the form of thin films.

3. Electric Properties of ZnO

3.1. Energy Band Gap

The electronic band structure of ZnO has been calculated by a number of groups [72-78]. Both the valence band maxima and the lowest conduction band minima occur at the Γ point k=0 [78] indicating that ZnO is a direct band gap semiconductor. The first two conduction band states are strongly Zn localized and correspond to empty Zn 3s levels. The higher conduction bands are free-electron-like. The O 2s bands associated with core-like energy states occur at around −20 eV. The band gap as determined from this calculation is 3.77 eV. This correlates reasonably well with the experimental value of 3.37 eV, and is much closer than the value obtained from standard LDA (local-density approximation) calculations, which tend to underestimate the band gap by ~3 eV due to its failure in accurately modeling the Zn 3d electrons.

In addition to calculations for the band structure of bulk ZnO, Ivanov and Pollmann have also carried out an extensive study on the electronic structure of the surfaces of wurtzitic ZnO [77]. Using the empirical tight-binding method (ETBM) to determine a Hamiltonian for the bulk states, the scattering theoretical method was applied to determine the nature of the surface states. The calculated data was found to be in very good agreement with experimental data obtained from electron-energy-loss spectroscopy (EELS) and ultra-violet photoelectron spectroscopy (UPS). Experimentally, the ZnO valence band is split into three band states, A, B and C by spin-orbit and crystal-field splitting.

The band gap has temperature dependence up to 300 K given by the relationship:

$$E_g(T) = E_g(T = 0)\frac{5.05\times10^{-4}T^2}{900-T} \tag{1}$$

3.2. Electrical Properties

Electrically, ZnO is a transparent-conducting-oxide semiconductor which at room temperature exhibits defect or impurity-dominated n-type conductivity, even in nominally undoped materials. This defect or impurity-dominated conductivity is a consequence of the large band gap [79-80] (≈3.37 eV at room temperature) combined with the unavoidable presence of electrically active native defects and impurities with donor ionization energies typically of ≈10–100 meV at concentrations of $\approx10^{15}$–10^{16} cm^{-3} [79]. Furthermore, while ZnO is readily doped n-type and carrier concentrations as high as $\approx10^{21}$ cm^{-3} are achievable [81], at present p-type doping remains challenging. Furthermore, in order to utilize ZnO-based materials for extending semiconductor optoelectronics further into the UV, two of the biggest technological challenges require to develop both synthesis and device processing techniques. The control on both methods would then achieve control over p-type doping as well as over metal/semiconductor junction formation (including both Schottky and ohmic contacts) to both n-type and p-type materials. Actually, most of what is known experimentally about the electronic properties of "pure" ZnO is actually based on n-type specimens.

To date, available data concerning fundamental electronic properties of intrinsic, accurately stoichiometric ZnO remains largely unknown [79]. In part this is because of the historical unattainable amount of sufficiently high-quality materials, a consequence of the difficulty (especially through bulk crystal-growth methods) in achieving adequate O incorporation into the specimen. Despite such a circumstance may seem merely a technological rather than a fundamental limitation, recent theoretical works [82-83] suggest that the n-type character results from electrically active and stoichiometric native point defects, such as vacancies, inter-stitials, and antisites [82]. In addition, the unusually non-amphoteric and donor-like character of unintentional but ever-present hydrogen [84] can cause the n-type property of ZnO. The n-type character may be inherent to ZnO, at least to ZnO materials produced through the use of near-equilibrium crystal growth methods.

In general, electrical transport properties of ZnO are directionally dependent due to the anisotropy of its wurtzite crystal structure. It is also of interest the electric-field dependence of the carrier mobility. A comparison [85] of the GaN and ZnO plots of drift velocity (v_d) versus electric field (μ) at low field reveals that $\mu_{ZnO} < \mu_{GaN}$ while at the same time the saturated drift velocities obey the opposite relation: $v_{ZnO} > v_{GaN}$. For hot-electron devices, where fields are high and transport is non-ohmic, it is v_{sat} rather than μ that can be the more important parameter to device operation [86].

For any practical implementation of ZnO-related materials as a viable electronic device technology, achieving control over p-type ZnO is of great importance. In addition, it will be fundamental the control over fabrication of high quality metal/semiconductor interfaces, including both ohmic contacts as well as Schottky barriers. Progresses between these two factors are in fact interdependent, since ohmic contacts ultimately will have to be developed for both n-type and p-type material layers. However, as the control over the ZnO material itself improves, similarly should happen with the control and reproducibility over the metal/semiconductor interface. Indeed, the control on the material quality of any metal is trivial if compared to that of the semiconductor, as well as the interface between them, as it was the case in the development of every technologically significant semiconductor/metal system.

Concerning the Schottky barriers, virtually all works reported to date were about n-type ZnO, due to the historical unavailability of p-type specimens. For example, the pioneering work of Mead [87] in the 1960s was on cleaved ZnO surfaces, as well as the flurry of recent works [88-95] were carried out on the best presently available (hydrothermally or pressurized-melt-grown) bulk single-crystal materials. Since ZnO is a wide-gap semiconductor with significant partially ionic bonding character, one might expect [96] Schottky barrier heights on n-type material (φ_{Bn}) to obey the simple Schottky–Mott model [97], that is

$$\varphi_{Bn}=\varphi_M-\chi_S \quad (2)$$

where φ_M is the work function of the metal and χ_S is the electron affinity of the semiconductor. Instead, most φ_{Bn} values for ZnO range from 0.6 to 0.8 eV regardless of metal used, with ideality factors near unity being more the exception than the rule [89-90]. Briefly, Schottky barriers were examined for numerous noble metals, including Ag and Au, [98-100] Pt [101-104] and Pd, [88, 96-102, 104-109] as well as metals such as Ni and Ir [89]. Allen et al. [90] have observed that only a small fraction of the available φ_{Bn} values for these metals from the literature involves "good" ideality factors, with in general a wide variability in reported barrier height values for any given metal. Part of the variability is due to surface polarity: for several metals, including Pt, Pd, and most recently [89-90] "silver oxide", substantially higher φ_{Bn} values were observed on Zn-polar than on O-polar {0001} surfaces [88-89]. This is explained in terms of surface polarity-dependent band bending due to the large spontaneous polarization of ZnO along its *c*-axis [90]. Another part of the variability was attributed to such factors as lateral inhomogeneity [88, 107]. This causes φ_{Bn} measured by I–V to be lower than that measured by C–V measurement, as well as a larger and also a linearly temperature-dependent ideality factor [107]. In addition, the most important part of the variability appears to be due to the influence of surface or near-surface defect states [89, 91-92]. For example, Allen and Durbin [89] have proposed that metal-induced gap states in the form of V_O, a known double donor with level approximately 0.7±0.2 eV below the conduction band [110], can form at the surface via metallurgical oxidation reaction of the metal (drawing O from the ZnO and thus forming V_O levels at enhanced concentration in a thin layer near the interface). The enhanced concentration of so formed V_O deep donors serves to pin the Fermi level (thereby causing departure from the Schottky–Mott model of Schottky barrier formation) and might account for the previously mentioned 0.6–0.8 eV range barrier heights irrespective of metal species, plus poor ideality factor, often observed on ZnO. To support this hypothesis, the same authors investigated φ_{Bn} for silver oxide (a species already reacted with oxygen thus almost inert), and found that, unusually, the silver oxide Schottky barriers exhibited an exceptionally low ideality factor of 1.06 for ZnO. Several other authors have also noted the beneficial effect on ideality factor by pre-treating the

ZnO surface prior to metallization with an oxygen-containing chemical [101] or plasma [92]: presumably, such treatments either remove adsorbates, or reduces the near-surface V_O, or possibly do both, with the specific phenomenological details varying from metal to metal, as well as depending upon inherent material quality of the ZnO specimen itself.

To apply ZnO nanostructures on various electronic devices, it is important to understand the nanostructure transport properties and their interaction with metal contacts. Lao et al. [111] have studied the transport properties of over 60 single circuits fabricated by aligning single ZnO nanobelts/nanowires across paired Au electrodes using dielectrophoresis. They found a spontaneous formation of an Au/ZnO nanowire Schottky diode in 80% of the samples when nanowire sizes were well controlled. This effect is likely due to the non-symmetric contacts at the two ends of the nanowires.

Concerning ohmic contacts, investigations on both n-type [103, 106, 112-115] and p-type [116-118] ZnO were performed. Generally speaking, to produce the best n-type ohmic contacts the most successful approach to date has been a multilayer metallization scheme, wherein the bottom layer (i.e. the layer in direct contact with ZnO) is reactive with nominally small φ_{Bn} to ZnO. In particular if that reaction tends to remove O from ZnO and the top layer is a noble metal the ohmic contact is further improved. For example, Kim et al. have investigated [114-115] Al/Pt contacts to n-type ($\approx 10^{18}$ cm^{-3}) ZnO, both as deposited, and, as a function of post-metallization anneal treatment. They obtained good specific contact resistance ($\approx 1x10^{-5}$ Ωcm^2) even without any annealing treatment. They attributed this effect to the Al/O interdiffusion resulting in the formation of a high density of near-surface V_O defects, thus constituting a double donor, as we noted earlier in our discussion on Schottky barriers. This effect is enhanced upon annealing at 300 °C for 1 min, whereas the specific contact resistance decreases [115] to $\approx 2\cdot 10^{-6}$ Ωcm^2. However, higher annealing temperature (up to 600 °C) is observed to be deleterious: evidently, an insulating layer of aluminum oxide compound (presumably Al_2O_3) forms at the interface. Therefore an insulator becomes interposed between the semiconductor and metal layers when the annealing time/temperature becomes too high, resulting in an increase (rather than a decrease) of the specific contact resistance. A somewhat analogous set of phenomenology is found in connection with a different metallization scheme: Ti/Al/Pt/Au [103, 106]. In this case, the annealing beyond even 200 °C is sufficient to cause a degradation [106]. Finally, additional metallization schemes were explored. Kim et al. examined Ti/Au contact [112-113] and Al/Pt [114-115]. Most recently these authors considered a Re/Ti/Au metallization scheme [119], wherein the latter proved to be metallurgically stable even following a high-temperature (700 °C, 1 min) annealing treatment as well as resulting in superior specific contact resistance of ≈$2x10^{-7}$ Ωcm^2 (for comparable ZnO doping levels) in comparison with Al/Pt. Studies on ohmic contacts to p-type ZnO are really few to date, but are becoming increasingly reported as p-type ZnO material availability and quality steadily improve. In 2005, the same Korean group (Kim et al.) have also studied the p-type material. In one study they employed a Ni/indium-tin oxide scheme [117] and in the other a Ni/Au scheme [116]. In both of them, Ni from the metallization interdiffused and reacted with the Zn from ZnO to form many intermetallic phases, ultimately yielding decent preliminary specific contact resistances (≈10^{-4} Ωcm^2) in each case [116-117]. Lastly, in a separate study on Au/Ni contacts by Mandalapu et al. [118] involving rapid thermal annealing (800 °C, 1 min), it was conjectured that the good specific contact resistance (again ≈10^{-4} Ωcm^2) was the result of Zn vacancies (due to Zn out diffusion) coupled with activated Sb atoms. Both effects seemed to increase the surface hole concentration which enables the formation of ohmic I/V characteristics at the metal/ZnO junction.

4. ZnO Nanowires Applications

4.1. ZnO Nanowire Based Electronic Devices

Semiconductor nanowires have drawn enormous research efforts for their special material properties and wide-ranging device applications with outstanding performances [120-122]. Among the II–VI semiconductors, zinc oxide (ZnO) nanowires were extensively studied for their abundant physical properties and versatile device applications. It is a polarized semiconductor that exhibits strong piezoelectric and pyroelectric properties. As stated in the previous sections, ZnO has a direct wide bandgap of 3.37 eV at room temperature. Compared with other wide bandgap materials, it has a larger exciton binding energy (60 meV) that ensures efficient exciton emissions at room temperature. In addition, due to the small size comparable to its Debye length, ZnO nanowire had shown superior sensitivity to chemical adsorbants. As a result, ZnO nanostructures have become a cutting edge of nanotechnology research in transparent electronics, actuators, sensors, optoelectronics, and spintronics. Three-terminal FET is a vital element in modern electronics. In order to construct electronic devices based on semiconductor nanowires, the fabrication of nanowire-based FET is the primary task. Three-terminal ZnO nanowire FET (NWFET) were fabricated and extensively explored in the past years. In the FET configuration, a pair of leads patterned onto the two ends of the nanowire channel serves as the source and drain electrodes. CMOS compatible photolithography and e-beam lithography are the frequently adopted techniques for electrode implementation. In addition to the source and drain electrodes, a weakly capacitively coupled terminal is employed to switch conduction on or off via the application of a transverse electric field [123-124]. This gate electrode provides a precise control over the channel conduction, and thus gates the FET device performances such as operation speed, ON/OFF ratio, and power efficiency.

Different gate structures were fabricated in order to examine the device operation. In the current progress of NWFET, there are four major types of gate configuration: back, top, side, and surrounding gates. The back gate configuration has been the traditional structure due to its fabrication simplicity; however, it lacks precision control of individual channel segments and poses difficulties for further device integration. In contrast, top, side, and surrounding gate structures involve a more complicated fabrication process, however they have demonstrated enhanced device characteristics and significantly increased integration density.

4.1.1. Back-Gated FET

By thermally growing a suitable thickness of SiO_2 layer on degenerately doped Si substrate, back-gate configured NWFET can be fabricated [125-127]. A single nanowire is placed on a Si back gate under a SiO_2 dielectric layer with source and drain electrodes attached to the two ends (Figure 11). The typical field-effect mobility of back-gated device falls in the range 3–80 $cm^2/V \cdot s$ with an ON/OFF ratio of around 10^4–10^6. Since the nanowire lies on the top of a flat dielectric substrate, the back gate capacitance can be estimated as:

$$C_{back} = (2\pi\varepsilon_r\varepsilon_0 L_{nw})/\ (\ln(2h/r_{nw})) \tag{3}$$

based on the cylinder-onplate model [59], where L_{nw} is the nanowire channel length, r_{nw} is the nanowire radius, h is the thickness of dielectric, and ε_r is the relative dielectric constant of the gate dielectric. The dielectric needs to be thin to achieve effective gate operation; however, thinner dielectric may result in significant leakage current and device failure [128]. To overcome this challenge, the use of high-k dielectric materials can facilitate the demand on thickness without inducing leakage current and causing dielectric breakdown.

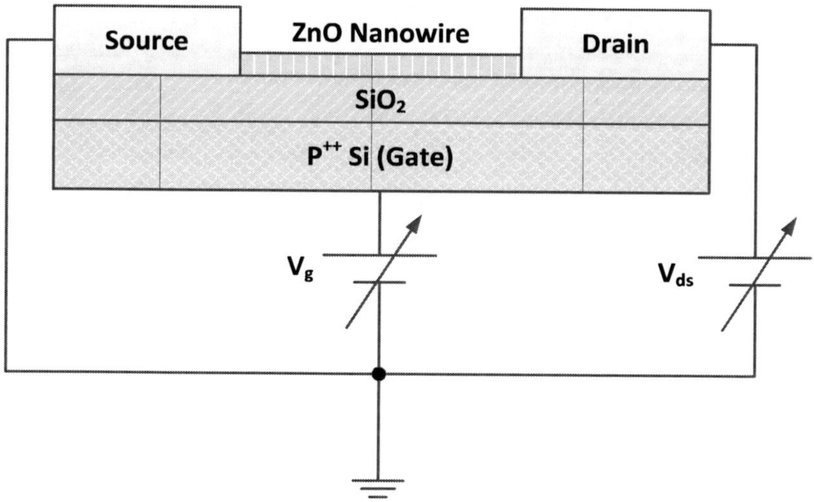

Figure 11. Back-gated ZnO NWFET on SiO_2/p^{++} Si substrate.

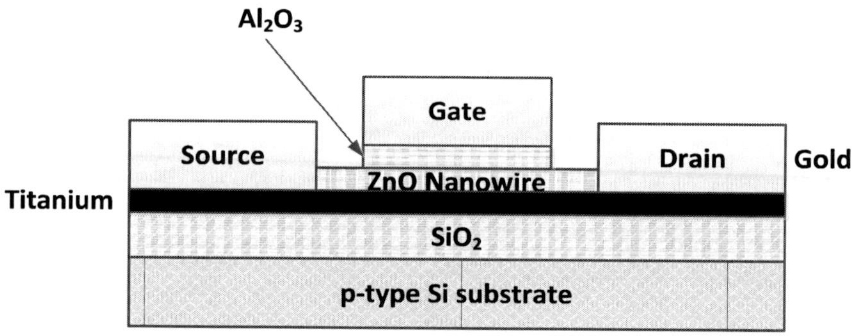

Figure 12. Top-gated ZnO NWFET patterned with 30 nm Ti and 100 nm Au as the source-drain electrodes and 20 nm Al_2O_3 as the dielectric gate.

4.1.2. Top-Gated FET

Although planar back gate configuration has been commonly adopted in NWFET fabrication, the top gate structure of NWFET [129] offers local gating response with precise control [130-131]. In addition, localized top gate electric field requires less field strength to efficiently regulate the channel conduction with an ON/OFF ratio reaching a value of 10^7. As shown in Figure 12, the source–drain contacts are first patterned, followed by a gate dielectric deposition. Keem et al. [131] used 100-nm Au as contact electrodes and made 20-nm-thick Al_2O_3 gate dielectric by atomic layer deposition (ALD). The gate dielectric layer forms an Ω shape, as shown in Figure 15b, to enhance the gate performance [132]. In such configuration, the top gate capacitance is expressed as [133]:

$$C_{top} = (2\pi\varepsilon_r\varepsilon_0 L_g)/\ln(r_g/r_{nw}) \tag{4}$$

where L_g is the length of nanowire under gate dielectric, r_{nw} is the nanowire diameter, and r_g is the outer diameter of surrounding dielectric layer. By using high-k material and adjusting the r_g/r_{nw} ratio, a large gate capacitance can be achieved, yielding improved transconductance and ON/OFF ratio over the common back gate configuration [132].

4.1.3. Side-Gated FET

In addition to back and top gates, side gating structure is also achieved in ZnO NWFET [134]. The metallic (normally Cr) gate electrodes are made just a few nanometers away from the suspended ZnO nanowire (Figure 13).

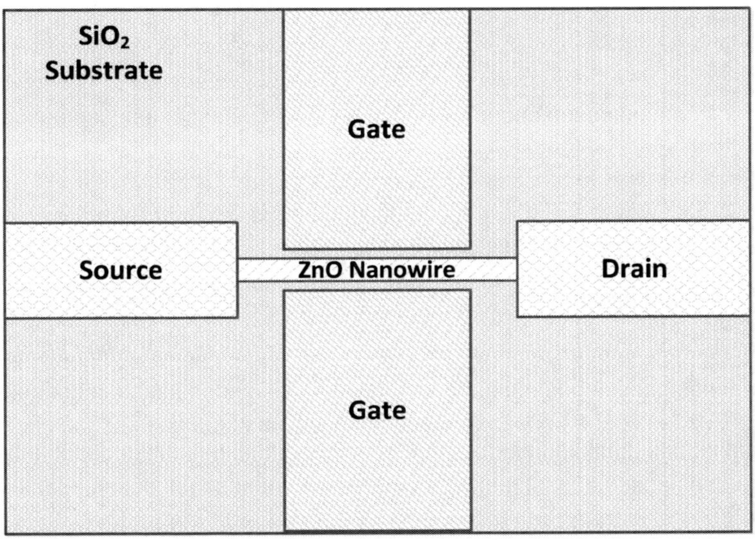

Figure 13. Top view of Side-Gated ZnO NWFET.

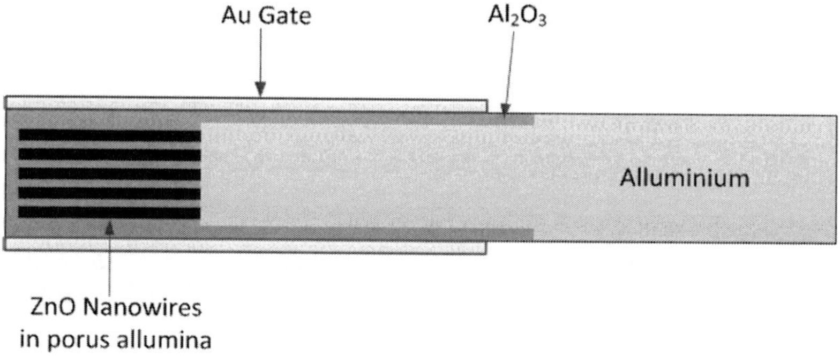

Figure 14. Vertically oriented ZnO nanowires for Vertical Sorrunding FET configuration.

With air as the dielectric gap, this side-gated device has shown field-effect mobility close to 1000 $cm^2/V \cdot s$, which is the highest reported value among non-passivated ZnO NWFETs. Its ON/OFF ratio is around 10^6. The advantage of this type of FET is that the gate insulator (air gap) is very small (~26 nm), leading to an excellent gate modulation efficiency. However, the difficulty of making an exact alignment of the nanowires among the electrodes limits the reproducibility in its fabrication.

4.1.4. Vertical Surrounding FET

In order to increase the device number per unit area and achieve ultrahigh integration density of FET devices, a lot of effort is devoted to explore vertically oriented NWFET [133]. With heavily p- or n-doped SiC as substrate, researchers at NASA Ames have successfully fabricated vertically aligned ZnO nanowires into p- and n-type vertical-surround-gate FETs (VSG-FET). The process is shown in Figure 14. ZnO nanowires are vertically grown on conducting SiC substrate. The surrounding gate dielectric (SiO_2) around the nanowire is uniformly coated by conformal CVD, followed by the deposition of the gate electrode (Cr). After planarization by chemical mechanical polishing, the top drain electrode (Cr) is subsequently patterned. The fabrication of VSG-FET requires strenuous procedures; however, the resulting VSG-FET compact structure features a 10% smaller footprint than the standard silicon-based MOSFET. There is an increasing interest in using porous AAO template matrix as the surrounding gate dielectric. As shown in Figure 14 an AAO template can be made into a rodlike structure with ZnO nanowires embedded in the template as the conduction channels [135]. Source–drain electrodes can be patterned at the two ends of the nanowires, and the outer shell (Au) functions as the gate electrode. The fabrication process of this type of surrounding gate FET is relatively simple and cost effective. Nanowire channel length and diameter are controllable by tailoring the AAO template. Although there are still unsolved problems in manufacturing large scale surrounding gate NWFETs, the success in fabricating vertical nanowire devices holds promising potential for Tera-level packing density of nanoelectronic devices.

Figure 15 shows a typical source–drain current versus gate voltage (I_{ds}–V_g) for as-grown ZnO NWFET. It is clear that positive V_g induces higher channel conductance as for typical n-type FET behavior. V_{gt} is the gate threshold voltage at which current is suppressed to an OFF state. The negative V_{gt} value indicates a depletion-mode FET behavior, i.e., the device is in the ON state at zero applied gate voltage. Based on the cylinder-on-plate model, field-effect mobility and carrier concentration of a back-gated NWFET can be calculated from the gate threshold voltage and transconductance:

$$g_m = \partial I_{ds}/\partial V_g \tag{5}$$

extracted from the linear region of I_{ds}–V_g. The carrier concentration induced by the gate V_g can be calculated as:

$$n = C_{back}|V_g - V_{gt}|/eAL_{nw} \tag{6}$$

where C_{back} is the back gate capacitance for a cylindrical wire of length L and cross-sectional area A. The field-effect mobility at fixed V_{ds} bias is [59]:

$$\mu = g_m L_{nw}^2/C_{back} V_{ds} \tag{7}$$

From the reported pristine ZnO NWFET without additional surface treatments, the effective carrier concentration is roughly estimated to be in the 10^{17}–10^{18} cm^{-3} range, the device ON/OFF ratio is around 10^4, and the mobility falls in the range from 3–80 $cm^2/V \cdot s$ [130, 136-138] which is comparable to those of ZnO thin film transistors [139-140].

It is highly desired to prepare 1D ZnO nanostructures doped with a selection of elements for enhancing and controlling their mechanical, electrical, and optical performance. So far, several doped 1D ZnO nanostructures were synthesized [141-146], being important for practical applications. Aluminum-doped zinc oxide (AZO), as a possible alternative to transparent conductive oxide (TCO), such as indium tin oxide (ITO), has attracted much attention because of its unique optical and electrical characteristics, as well as its low cost and non-toxicity. Lin et. al.

[147] have fabricated a photoconductor device based on a single AZO nanowire that is sensitive to illumination with below-gap light and with an assembled memory. Well-aligned and highly ordered architectures of 1D nanostructures were thus required and fabricated by electrospinning.

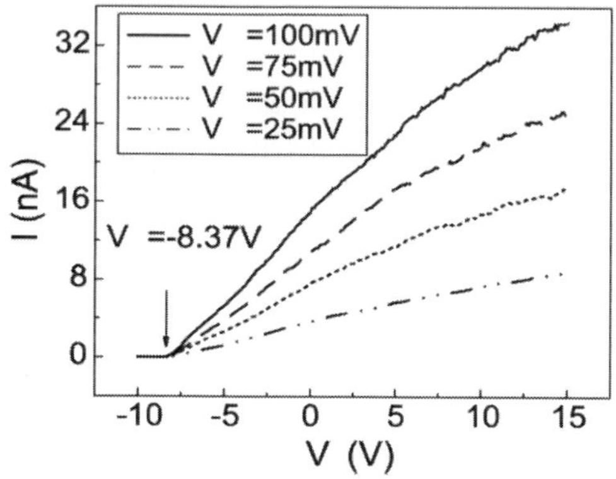

Figure 15. I_{ds}–V_g curves show depletion-mode behavior with a threshold voltage at −8.37 V. For this NWFET (60 nm in diameter), the carrier concentration is estimated to be ~1.4 × 1017 cm^{-3}, and mobility is 17.2 $cm^2/V \cdot s$.

The nanowires were then directly transferred to a silicon substrate with a 500 nm-thick thermally grown SiO_2 layer and calcined at 550 °C at a heating rate of 10 °C min^{-1} in air for 3 h to obtain well-aligned polycrystalline AZO nanowires. Gold electrodes were then deposited on the AZO nanowires by thermal evaporation using a mask to form channels that were 500 nm in length.

4.2. ZnO Nanowire-Based Sensor Devices

Due to their amazing electronic, optical, and biocompatible properties for sensing, ZnO nanowires have been attracting great deal of attention for detecting a number of different physical phenomena such as chemical and biological sensors [148], pH sensors [149-151], gas sensors [152-153], mechanical sensors [154-155], photo detectors [156-157], photovoltaic applications or temperature sensors [158].

4.2.1. Gas Sensors

The electrical conductivity of ZnO can be dramatically affected by the adsorption and desorption of gas species on their surface [159]. In particular 1D ZnO nanostructures, with their high aspect ratio and thus surface area, give rise to a high sensibility and selectivity with respect to gas molecules. The adsorption of gas molecules on the surface of ZnO induces the change of the electrical properties which can be thus measured as shown previously [160-161]. Normally, in ambient air, some oxygen molecules are adsorbed on the surface of the naturally n-type ZnO and subsequently reduced to oxygen ions (O_2^-, O^-, and O^{2-}) by capturing electrons from the ZnO conduction band. This adsorption results in an increased electrical resistance [162]. When exposed to a reducing gas such as ethanol (CH_2OH), acetone (C_3H_6O), ammonia (NH_3), carbon monoxide (CO), and hydrogen (H_2) at moderate temperatures, the gas reacts with the adsorbed oxygen ions, releasing the trapped electrons back to the conduction band, and thus leading to an increased carrier concentration and increased electrical conduction [161] Wang et al. [159] reported on ZnO-based gas sensors using hydrothermally grown ZnO nanowire arrays. They found an excellent sensitivity to H_2, NH_3 and CO. In addition, the surface modification by chemical functionalizing agents can improve the sensitivity of ZnO as gas sensors. For example, Kakati et al. reported on ZnO nanowires with a indium antimonide (InSb) surface modification, finding that the acetone sensitivity was significantly improved due to the reduced activation energy. [162]

4.2.2. pH Sensors

The surface electrical charge density of ZnO change with pH in electrolyte [149, 163]. The pH response of ZnO, as well as for in general all metal oxide surfaces, resides in the formation of hydroxyl groups depending on the solution pH. This leads to a pH-dependent net surface change with a resulting change in voltage drop at the semiconductor/liquid interface [164-165]. Based on this principle, ZnO nanorods were used to fabricate a highly sensitive pH sensor on Ag-coated borosilicate glass Femtotio® II capillaries to detect the intracellular pH of a human fat cell [149]. Other authors [166] have shown pH-sensing devices based on single ZnO nanorods with Ohmic contacts at either ends, exhibiting large changes in current upon exposing the surface region to polar liquid electrolytes. The polar species led to a change of surface charges on the ZnO nanorod, producing a change in surface potential at the semiconductor/liquid interface, with a linear change in conductance between pH 2 and 12 of 8.5 nS/pH in the dark with a stable operation and a resolution of 0.1 pH over the entire pH range.

Other authors have reported on the use of single ZnO microwires among nanogap-sized gold electrodes for the *in-situ* measurement of the pH solution. In particular, the authors have exploited the current variation of the ZnO-gold junction, leading to Schottky contacts [167] or to an enhanced gate current variation using the ZnO as NWFET [151] using aminepropyl-functionalized ZnO microwires. In both cases an increase of the detected current was obtained at acidic pH starting from neutral solution. In case of basic pH, a strong decrease of the current variation on the NWFET device was proved.

4.2.3. UV Sensors

ZnO has also shown great potentials for optical applications, such as UV detection. The high sensitivity of ZnO nanowires to UV light is due to the presence of oxygen-related hole trapping states at their surface, as proposed by Soci et al. [168]. Interestingly, the UV sensitivity of a single ZnO nanowire can be enhanced by applying a mechanical strain to it [169]. The authors in particular applied different bending strains and excitation light intensities to metal-semiconductor (ZnO)-metal junctions, forming a Schottky barrier. The resulting photoconducting responses were then evaluated, thus exploiting the coupling of optical, mechanical, and electrical properties of ZnO nanowires.

4.3. Piezoelectric Nanogenerators

The piezoelectricity of zinc oxide is related to its non-centrosymmetric crystal structure. The zinc atom, which has positive charge, is tetrahedrally bonded to the oxygen atoms which are negatively charged, resulting in both charge of the center of gravity located in the crystal structure [4]. Under the application of a lattice deformation, the positive and negative charge centers of gravity will no longer coincide. This crystal deformation will then induce a formation of a local electric dipole inside the cell. In the case that all the cells have the same orientation and thus all the local electric dipoles are aligned, a macroscopic electric dipole inside the crystal is generated. Therefore piezoelectric materials are able to convert mechanical deformation in electrical signal. Moreover piezoelectric effect is also reversible, thus when the material is subjected to an electric field the ions in the cell moved inducing a deformation of the whole crystal [4].

Combining the piezoelectric properties with the semiconducting ones, zinc oxide nanowires were used as nanogenerators to harvest energy from the surrounding environment, thus by converting mechanical deformations into electric potential. The quantity of converted energy is normally small, of the order of μW up to few mW, but it is suitable for the charging of mobile phone or wearable electronic which have become extremely diffuse in nowadays society. The goal of nanogenerators is indeed to create portable harvester devices to solve the problem of the accessibility to the energy power sources in everyday life [4, 170].

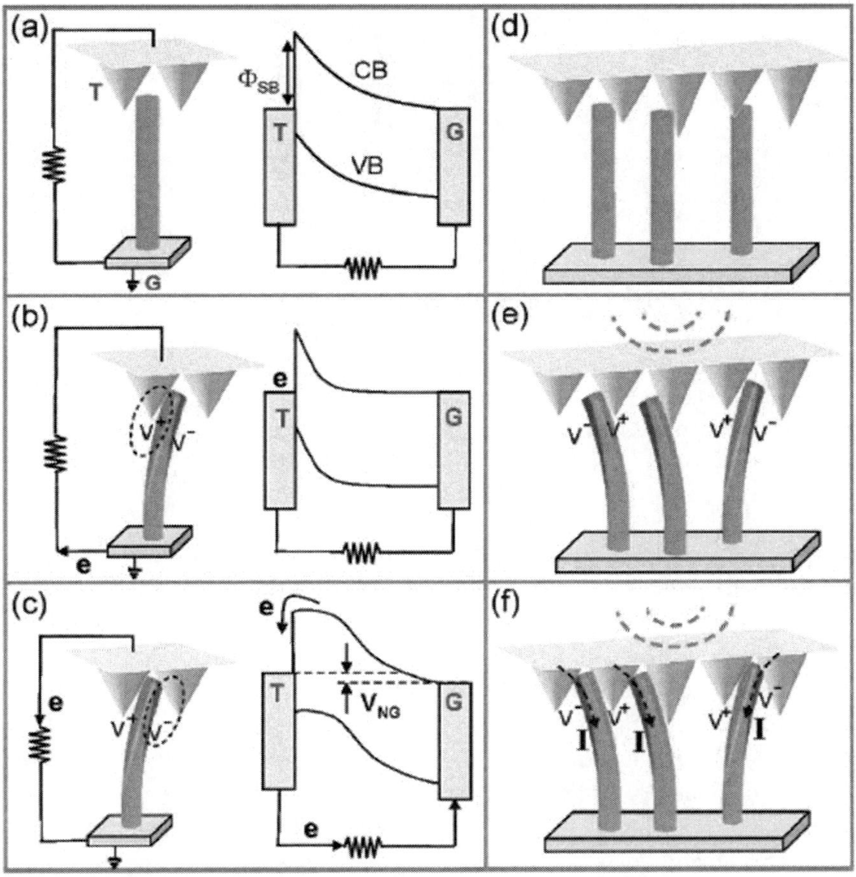

Figure 16. Physical principle of ZnO NWs nanogenerators exploiting zigzag electrode under the excitation of an ultrasonic wave. (a, d) When no deformation is induced, the nanowires form a Schottky barrier with the zigzag electrode and no current is flowing in the circuit. (b, e) Under the excitation of an ultrasonic wave, the nanowires are deformed and the electrode tips first contact the elongated part with a positive local potential. The barrier is then reversely biased and no current is flowing. (c, f) The movement of the tips increased under the action of the ultrasonic wave and they contact the compressed part of the NWs with a negative local potential. The barrier is then forwardly biased and the electrons can flow from the NWs into the tip and the external circuit. Reprinted with permission from [172]. Copyright (2008) American Chemical Society.

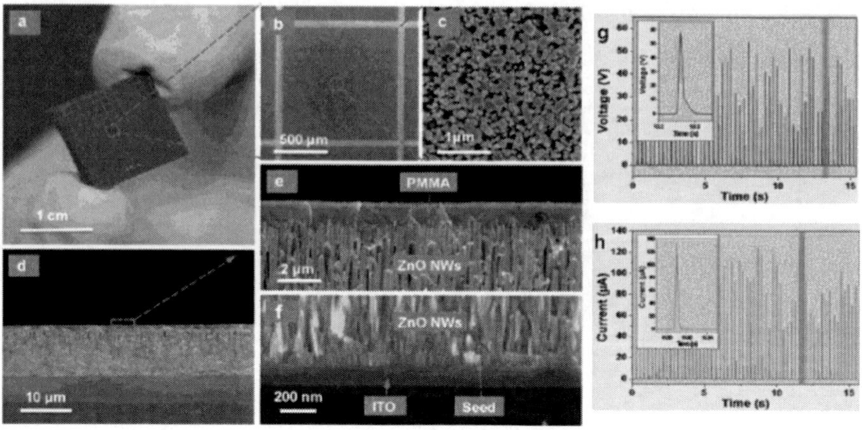

Figure 17. (a-f) Structure of the nanogenerator showing ZnO nanowires surrounded by a thin layer of PMMA. (g-h) Open-circuit voltage and close-circuit current generated by the nanogenerator connected to a bridge rectifier under the impact of a human palm. Adapted with permission from [177]. Copyright (2012) American Chemical Society.

The nanogenerator concept was firstly introduced by Professor Zhong Lin Wang in 2006 analyzing the electrical potential generation of an individual zinc oxide nanowires when deformed by means of a platinum-coated Atomic Force Microscopy (AFM) tip [5]. Since a Schottky barrier was formed between the tip and the nanowire, when the tip contacted the elongated part of the nanowires (local positive potential), no electric potential was generated because of the reversed biased Schottky behavior. In contrast, when the tip was about to leave the nanowires and thus in contact with the compressed part (local negative potential), the Schottky contact was forward biased and thus a sharp voltage peak was generated. The output power of a single ZnO NWs was estimated of ~0.5pW.

The first example of nanogenerator devices was again proposed by Wang substituting the AFM tip with a zigzag electrode placed on the top of an array of vertical ZnO NWs [171-172]. In this way, under the application of a mechanical stimulus by means of an ultrasonic wave, all the NWs were deformed simultaneously (Figure 16) and the devices were able to generate an almost constant current output of ~500 nA and voltage output of ~10 mV, resulting in an electrical power density of ~83 nW/cm^2.

The advantage of these nanogenerators based on ZnO NWs, with respect to the classical mechanical harvester, is that they are not based on mechanical resonance and thus could work in a wide range of frequency and therefore are independent from the final application [4].

From these two starting works, several nanogenerator devices based on ZnO NWs were developed in the last years in order to further increase the generated electrical power. 2-D nanogenerators were developed by growing nanowires on a flexible substrate between two electrodes [173]. The electrical output was then generated by bending the device. Nanowires were also grown on fibers in order to fabricate wearable fabric nanogenerators which convert mechanical deformation induced by bending, stretching, sliding or brushing between the composite fibers [174-175].

The highest output was reached with composite nanogenerators having a dielectric polymer as matrix. In these devices the ZnO NWs were completely covered or dispersed in a polymer or just separated from the top electrode by a thin layer of insulating material [175-177]. In the latter case, the Schottky contact of the first device configurations was replaced with the insulating layer which prevents the current leakage between ZnO-metal interface. Moreover the polymer capping layer better distributes the compressive force to all the NWs, independently from their length differences, and also improves the robustness of the nanogenerator device (Figure 17). Thanks to these improvements it was possible to reach a current output of ~134 μA and voltage output of ~58 V, resulting in an instantaneous electrical power density of ~0.78 mW/cm^2 [177].

Even if it was possible to reach very preforming harvest devices using ZnO nanowires, the research in these kind of nanogenerator family was drastically reduced in the last year when the first triboelectric nanogenerators were presented in January 2012 [178]. These devices convert the mechanical energy into electrical ones exploiting both triboelectric effect and electrostatic induction between two thin organic/inorganic films that exhibit opposite tribo-polarity [170]. Actually the best performing triboelectric nanogenerator could generate current output of ~2 mA and voltage output of ~1200 V, resulting in an instantaneous electrical power density of ~0.313 W/cm^2, sufficient to light able up to 600 commercial LED with one compression made with human palm [179].

4.4. Photoelectrochemical Cells

Particular interest is given to the development of clean and renewable energy sources such as the conversion of solar energy into electrical and chemical energy throughout photoelectrochemical cells (PEC). In general, TiO_2 has shown good performances as photoanode for both dye sensitized solar cells (DSSC) [180] and water splitting (WS) systems [181].

ZnO has already resulted to be an interesting metal oxide to compete with TiO_2. Zinc oxide has approximately the same energy band gap and electron affinity as titanium dioxide, and ZnO DSSCs built using ZnO nanoparticles have shown the second highest efficiencies (up to 6.5%) after TiO_2 [182]. ZnO possesses also other critical criteria for water splitting such as an appropriate flat band potential, low electrical resistance, and good corrosion resistance in aqueous solution [183].

In particular, the 1D configuration of nanowires gives to ZnO extra advantages for the use as photoelectrode for the production of solar energy. The electron diffusion coefficient of ZnO NWs is several hundred times larger than that of ZnO or TiO_2 nanoparticle-based films [184]. With aligned ZnO nanowires several problems can be solved in devices based on the nanoporous materials, which rely on the trap-limited diffusion for electron transport, a slow mechanism that limits electron extraction efficiency. These structures can be used directly as photoelectrodes with dye molecules adsorbed on their surface [185]. To overcome the low surface area that 1D-nanostructures have compared with nanoparticle-based films, some studies combined both structures in order to take advantage of the benefits of each of them: improved electron transport along the vertically aligned nanowire array and increased surface area provided by the nanoparticles. An example of hybrid structures is the combination of vertically aligned ZnO nanowires with TiO_2 nanoparticles anchored top their surface [182, 186].

Recent studies have shown that N or Al doped ZnO generates a significant red shift in light absorption wavelength [187-188], resulting in a higher incident-photon-to-current-conversion efficiency (IPCE) value. Thus, ZnO-based PECs were fabricated showing a high overall efficiency.

However, one of the new challenges for the scientists is to create ZnO based DSSCs and WS cells with improved efficiency. In particular, reducing the recombination at the photoanode electrolyte interface with closely spaced nanowires, approximately twice the exciton diffusion length (<40 nm) and covering the nanowires with nanoparticles to increase the surface area, is expected to enhance the power conversion efficiency [189].

Conclusion

In the present chapter we have shown different synthesis techniques of ZnO nanowires and their applications related to the electrical properties. It is clear that ZnO is a very versatile material and the nanowire morphology can be obtained through various and different synthesis methods. Its electrical properties allows the use of ZnO nanowires as semiconductor for the construction of Field Effect Transistors devices. In addition the combination of both semiconductor and piezoelectric character of ZnO give rise to a series of interesting application in the field of both sensors and nanogenerators. These unique properties, as well as the photogeneration ability, the biocompatibility of ZnO and many other characteristics, makes ZnO nanowires and in general nanostructures as fundamental building block to be integrated in new and miniaturized smart devices, such as micro-electro-mechanical systems (MEMS) and read-out electronic circuits. The combination of different properties in one device, such as the piezotronic and the piezo-phototronic effects, the gas or pH sensing ability, would lead to multifunctional but extremely compact nanodevices.

References

[1] Morkoç, H.; Özgür, Ü., Zinc Oxide: Fundamentals, Materials and Device Technology. Wiley-WCH: 2009.

[2] Gomez, J.; Tigli, O., Zinc oxide nanostructures: from growth to application. *J. Mater. Sci.* 2013, 48 (2), 612-624.

[3] Wang, Z. L., Zinc oxide nanostructures: growth, properties and applications. *J. Phys.: Condens. Matter.* 2004, 16, R829-R858.

[4] Wang, Z. L., ZnO nanowire and nanobelt platform for nanotechnology. *Materials Science and Engineering: R: Reports* 2009, 64 (3–4), 33-71.

[5] Wang, Z. L.; Song, J., Piezoelectric Nanogenerators Based on Zinc Oxide Nanowire Arrays. *Science* 2006, 312, 242-246.

[6] Ozgur, U.; Alivov, Y. I.; Liu, C.; Teke, A.; Reshchikov, M. A.; Dogan, S.; Avrutin, V.; Cho, S.-J.; Morkoc, H., A comprehensive review of ZnO materials and devices. *J. Appl. Phys.* 2005, 98 (4), 041301.

[7] Tian, Z. R.; Voigt, J. A.; Liu, J.; McKenzie, B.; McDermott, M. J.; Rodriguez, M. A.; Konishi, H.; Xu, H., Complex and oriented ZnO nanostructures. *Nat. Mater.* 2003, 2 (12), 821-826.

[8] Xu, S.; Wang, Z. L., One-Dimensional ZnO Nanostructures: Solution Growth and Functional Properties. *Nano Res.* 2011, 4 (11), 1013-1098.

[9] Xu, C.; Shin, P.; Cao, L.; Gao, D., Preferential Growth of Long ZnO Nanowire Array and Its Application in Dye-Sensitized Solar Cells. *J. Phys. Chem.* C 2009, 114 (1), 125-129.

[10] Weintraub, B.; Zhou, Z.; Li, Y.; Deng, Y., Solution synthesis of one-dimensional ZnO nanomaterials and their applications. *Nanoscale* 2010, 2, 1573-1587.

[11] Pacholski, C.; Kornowski, A.; Weller, H., Self-Assembly of ZnO: From Nanodots to Nanorods. *Angew. Chem. Int. Ed.* 2002, 41 (7), 1188-1191.

[12] Öztürk, S.; Tasaltın, N.; Kılınç, N.; Yüzer, H.; Öztürk, Z. Z., Fabrication of ZnO nanowires at room temperature by cathodically induced sol–gel method. *Appl. Phys.* A 2010, 99, 73-78.

[13] Ramirez, D.; Pauporte, T.; Gomez, H.; Lincot, D., Electrochemical growth of ZnO nanowires inside nanoporous alumina templates. A comparison with metallic Zn nanowires growth. *Phys. Status Solidi A* 2008, 205 (10), 2371-2375.

[14] Bechelany, M.; Amin, A.; Brioude, A.; Cornu, D.; Miele, P., ZnO nanotubes by template-assisted sol–gel route. *J. Nanopart. Res.* 2012, 14, 980-987.

[15] Huang, M. H.; Wu, Y.; Feick, H.; Tran, N.; Weber, E.; Yang, P., Catalytic Growth of Zinc Oxide Nanowires by Vapor Transport. *Adv. Mater.* 2001, 13 (2), 113-116.

[16] Zhu, Z.; Chen, T.-L.; Gu, Y.; Warren, J.; Osgood, R. M., Zinc Oxide Nanowires Grown by Vapor-Phase Transport Using Selected Metal Catalysts: A Comparative Study. *Chemistry of Materials* 2005, 17 (16), 4227-4234.

[17] Yu, D.; Trad, T.; McLeskey, J.; Craciun, V.; Taylor, C., ZnO Nanowires Synthesized by Vapor Phase Transport Deposition on Transparent Oxide Substrates. *Nanoscale Res. Lett.* 2010, 5 (8), 1333 - 1339.

[18] Wu, J. J.; Liu, S. C., Catalyst-Free Growth and Characterization of ZnO Nanorods. *J. Phys. Chem.* B 2002, 106, 9546.

[19] Park, W. I.; Kim, D. H.; Jung, S. W.; Yi, G. C., Metalorganic vapor-phase epitaxial growth of vertically well-aligned ZnO nanorods. . *Appl. Phys. Lett.* 2002, 80, 4232–4234.

[20] Sun, Y.; Fuge, G. M.; Ashfold, M. N. R., Growth of aligned ZnO nanorod arrays by catalyst-free pulsed laser deposition methods. . *Chem. Phys. Lett.* 2004, 396, 21–26.

[21] Chiou, W. T.; Wu, W. Y.; Ting, J. M., Growth of single crystal ZnO nanowires using sputter deposition. . *Diam. Relat. Mater.* 2003, 12, 1841–1844.

[22] Govender, K.; Boyle, D. S.; Kenway, P. B.; O'Brien, P., Understanding the factors that govern the deposition and morphology of thin films of ZnO from aqueous solution. *J. Mater. Chem.* 2004, 14, 2575–2591.

[23] Chen, P.; Gu, L.; Xue, X.; Song, Y.; Zhua, L.; Cao, X., Facile synthesis of highly uniform ZnO multipods as the supports of Au and Ag nanoparticles. *Mater. Chem. Phys.* 2010, 122, 41-48.

[24] Jiang, P.; Zhou, J.-J.; Fang, H.-F.; Wang, C.-Y.; Wang, Z. L.; Xie, S.-S., Hierarchical Shelled ZnO Structures Made of Bunched Nanowire Arrays. *Adv. Funct. Mater.* 2007, 17, 1303-1310.

[25] Rivera, V. F.; Auras, F.; Motto, P.; Stassi, S.; Canavese, G.; Celasco, E.; Bein, T.; Onida, B.; Cauda, V., Length-dependent charge generation from vertical arrays of high-aspect ratio ZnO nanowires. *Chem. Eur. J.* 2013, doi: 10.1002/chem.201204429.

[26] Zhitao, H.; Sisi, L.; Junjun, L.; Jinkui, C.; Yong, C., Facile synthesis of ZnO nanowires on FTO glass for dye-sensitized solar cells. *J. Semiconductors* 2013, 34 (7), 074002.

[27] Greene, L. E.; Law, M.; Tan, D. H.; Montano, M.; Goldberger, J.; Somorjai, G.; Yang, P., General Route to Vertical ZnO Nanowire Arrays Using Textured ZnO Seeds. *Nano Lett.* 2005, 5 (7), 1231-1236.

[28] Liu, J. P.; Xu, C. X.; Zhu, G. P.; Li, X.; Cui, Y. P.; Yang, Y.; Sun, X. W., Hydrothermally grown ZnO nanorods on self-source substrate and their field emission. *J. Phys. D, Appl. Phys.* 2007, 40 (7), 1906.

[29] Laurenti, M.; Cauda, V.; Gazia, R.; Fontana, M.; Rivera, V. F.; Bianco, S.; Canavese, G., Wettability control on ZnO nanowires driven by seed layer properties. *Eur. J. Inorg. Chem.* 2013, 2013 (14), 2520-2527.

[30] Podrezova, L. V.; Cauda, V.; Cicero, G.; Fontana, M.; Porro, S., Comparison between ZnO nanowires grown by chemical vapour deposition and hydrothermal synthesis. *Appl. Phys* A 2013, 113, 623-632.

[31] Elias, J.; Tena-Zaera, R.; Lévy-Clément, C., Electrochemical deposition of ZnO nanowire arrays with tailored dimensions. *Journal of Electroanalytical Chemistry* 2008, 621 (2), 171-177.

[32] Zheng, M.; Zhang, L.; Li, G.; Shen, W., Fabrication and optical properties of large-scale uniform zinc oxide nanowire arrays by one-step electrochemical deposition technique. *Chemical Physics Letters* 2002, 363 (1), 123-128.

[33] Prună, A.; Pullini, D.; Mataix, D. B., Influence of deposition potential on structure of ZnO nanowires synthesized in track-etched membranes. *Journal of The Electrochemical Society* 2012, 159 (4), E92-E98.

[34] Postels, B.; Bakin, A.; Wehmann, H.-H.; Suleiman, M.; Weimann, T.; Hinze, P.; Waag, A., Electrodeposition of ZnO nanorods for device application. *Applied Physics A* 2008, 91 (4), 595-599.

[35] Xu, L.; Guo, Y.; Liao, Q.; Zhang, J.; Xu, D., Morphological Control of ZnO Nanostructures by Electrodeposition. *The Journal of Physical Chemistry* B 2005, 109 (28), 13519-13522.

[36] A. Kathalingam, M. R. K., Y. S. Chae and J. K. Rhee, Studies on Electrochemically Deposited ZnO Thin Films *Journal of the Korean Physical Society* 2009, 55 (6), 2476 - 2481.

[37] Öztürk, S.; Taşaltın, N.; Kılınç, N.; Yüzer, H.; Öztürk, Z. Z., Fabrication of ZnO nanowires at room temperature by cathodically induced sol–gel method. *Applied Physics A* 2010, 99 (1), 73-78.

[38] Wang, Z.; Li, H., Highly ordered zinc oxide nanotubules synthesized within the anodic aluminum oxide template. *Applied Physics A* 2002, 74 (2), 201-203.

[39] Coasne, B.; Mezy, A.; Pellenq, R. J. M.; Ravot, D.; Tedenac, J. C., Zinc Oxide Nanostructures Confined in Porous Silicas. *Journal of the American Chemical Society* 2009, 131 (6), 2185-2198.

[40] Cao, G.; Liu, D., Template-based synthesis of nanorod, nanowire, and nanotube arrays. *Advances in colloid and interface science* 2008, 136 (1), 45-64.

[41] Sulka, G.; Stroobants, S.; Moshchalkov, V.; Borghs, G.; Celis, J.-P., Synthesis of well-ordered nanopores by anodizing aluminum foils in sulfuric acid. *Journal of The Electrochemical Society* 2002, 149 (7), D97-D103.

[42] Li, L.; Pan, S.; Dou, X.; Zhu, Y.; Huang, X.; Yang, Y.; Li, G.; Zhang, L., Direct electrodeposition of ZnO nanotube arrays in anodic alumina membranes. *The Journal of Physical Chemistry C* 2007, 111 (20), 7288-7291.

[43] Ottone, C.; Bejtka, K.; Chiodoni, A.; Farías, V.; Canavese, G.; Stassi, S.; Cauda, V., Comprehensive study of the templating effect on the ZnO nanostructure formation within porous hard membranes. *New J. Chem.* 2014, 10.1039/C3NJ01135F.

[44] Pierson, H. O., Handbook of Chemical Vapor Deposition (CVD), Principles, Technology, and Applications. Second Edition ed.; William Andrew Publishing, LLC: Norwich, New York 1999.

[45] Wu, C. L.; Chang, L.; Chen, H. G.; Lin, C. W.; Chang, T. F.; Chao, Y. C.; Yan, J. K., Growth and characterization of chemical-vapor-deposited zinc oxide nanorods. *Thin Solid Films* 2006, 498, 137-141.

[46] Wei, M.; Zhi, D.; MacManus-Driscoll, J. L., Self-catalysed growth of zinc oxide nanowires. *Nanotechnology* 2005, 16, 1364-1368.

[47] Chang, P.-C.; Fan, Z.; Wang, D.; Tseng, W.-Y.; Chiou, W.-A.; Hong, J.; Lu, J. G., ZnO nanowires synthesized by vapor trapping CVD method. *Chem. Mater.* 2004, 16, 5133-5137.

[48] Cha, S. N.; Song, B. G.; Jung, J. E.; Han, I. T.; Ha, J. H.; Hong, J. P.; Kang, D. J.; Kim, J. M., Controlled growth of vertically aligned ZnO nanowires with different crystal orientation of the ZnO seed layer. *Nanotechnology* 2008, 19, 235601.

[49] Li, C.; Fang, G.; Li, J.; Ai, L.; Dong, B.; Zhao, X., Effect of seed layer on structural properties of ZnO nanorod arrays grown by vapor-phase transport. *J. Phys. Chem. C* 2008, 112, 990-995.

[50] Das, S. N.; Kar, J. P.; Xiong, J.; Myoung, J.-M., Syntehsis of ZnO Nanowire by MOCVD technique: effect of substrate and growth parameter. InTech Open: 2012.

[51] Juàrez, J. M.; Andraca, J. A.; Cuenca, R.; Jaramillo, D.; Juàrez, F., MOCVD growth of ZnO nanowires through colloidal and sputtered Au seed via $Zn[TMHD]_2$ precursor. *Physics Procedia E* 2012, 25, 278-286.

[52] Kim, S.-W.; Fujita, S.; Fujita, S., ZnO nanowires with high aspect ratios grown by metalorganic chemical vapor deposition using gold nanoparticles. *Appl. Phys. Lett.* 2005, 86, 153119-3.

[53] Xu, S.; Lao, C.; Weintraub, B.; Wang, Z. L., Density-controlled growth of aligned ZnO nanowire arrays by seedless chemical approach on smooth surfaces. *J. Mater. Res.* 2008, 23.

[54] Lee, W.; Jeong, M.-C.; Myoung, J.-M., Catalyst-free growth of ZnO nanowires by metal-organic chemical vapor deposition (MOCVD) and thermal evaporation. *Acta Materialia* 2004, 52, 3949-3957.

[55] Black, K.; Chalker, P. R.; Jones, A. C.; King, P. J.; Roberts, J. L.; Heys, P. N., A new method for the growth of zinc oxide nanowires by MOCVD using oxygen-donor adducts of dimethylzinc. *Chem. Vap. Deposition* 2010, 16, 106-111.

[56] Valerini, D.; Caricato, A. P.; Lomascolo, M.; Romano, F.; Taurino, A.; Tunno, T.; Martino, M., Zinc oxide nanostructures grown by pulsed laser deposition. *Appl. Phys. A* 2008, 93, 729-733.

[57] Franklin, J. B.; Zou, B.; Petrov, P.; McComb, D. W.; Ryan, M. P.; Lachlan, M. A. M., Optimised pulsed laser deposition of ZnO thin films on transparent conducting substrates. *J. Mater. Chem.* 2011, 21, 8178-8182.

[58] Lemlikchi, S.; Abdelli-Messaci, S.; Lafane, S.; Kerdja, T.; Guittoum, A.; Saad, M., Study of structural and optical properties of ZnO films grown by pulsed laser deposition. *Appl. Surf. Sci.* 2010, 256, 5650-5655.

[59] Cao, B. Q.; Lorenz, M.; Rahm, A.; Wenckstern, H. v.; Czekalla, C.; Lenzner, J.; Benndorf, G.; Grundmann, M., Phosphorus acceptor doped ZnO nanowires prepared by pulsed-laser deposition. *Nanotechnology* 2007, 18, 455707-5.

[60] Rahm, A.; Lorenz, M.; Nobis, T.; Zimmermann, G.; Grundmann, M.; Fuhrmann, B.; Syrowatka, F., Pulsed-laser deposition and characterization of ZnO nanowires with regular lateral arrangement. *Appl. Phys. A* 2007, 88, 31-34.

[61] Liu, Z. W.; Ong, C. K.; Yu, T.; Shen, Z. X., Catalyst-free pulsed-laser-deposited ZnO nanorods and their room-temperature photoluminescence properties. *Appl. Phys. Lett.* 2006, 88, 053110-3.

[62] Hartanto, A. B.; Ning, X.; Nakata, Y.; Okada, T., Growth mechanism of ZnO nanorods from nanoparticles formed in a laser ablation plume. *Appl. Phys. A* 2004, 78, 299-301.

[63] Yang, B.; Kumar, A.; Zhang, H.; Feng, P.; Katiyar, R. S.; Wang, Z., Growth of ZnO nanostructures on metallic and semiconducting substrates by pulsed laser deposition technique. *J. Phys. D Appl. Phys.* 2009, 42, 045415-7.

[64] Koidis, C.; Logothetidis, S.; Georgiou, D.; Laskarakis, A., In-situ and real-time investigation of ZnO thin films growth onto rigid and flexible substrates. *Phis. Stat. Sol. C* 2008, 5, 1366-1369.

[65] Dang, W. L.; Fu, Y. Q.; Luo, J. K.; Flewitt, A. J.; Milne, W. I., Deposition and characterization of sputtered ZnO films. *Superlattice Microst.* 2007, 42, 89-93.

[66] Pan, C. T.; Liu, Z. H.; Chen, Y. C.; Liu, C. F., Design and fabrication of flexible piezo-microgenerator by depositing ZnO thin films on PET substrates. *Sens. Actuat. A* 2010, 159, 96-104.

[67] Choi, D.; Lee, K. Y.; Lee, K. H.; Kim, E. S.; Kim, T. S.; Lee, S. Y.; Kim, S.-W.; Choi, J.-Y.; Kim, J. M., Piezoelectric touch-sensitive flexible hybrid energy harvesting nanoarchitectures. *Nanotechnology* 2010, 21, 405503-6.

[68] Kuoni, A.; Holzherr, R.; Boillat, M.; Rooij, N. F. d., Polyimide membrane with ZnO piezoelectric thin film pressure transducers as a differential pressure liquid flow sensor. *J. Micromech. Microeng.* 2003, 13, S103-S107.

[69] Kim, S.; Jeong, M.-C.; Oh, B.-Y.; Lee, W.; Myoung, J.-M., Fabrication of Zn/ZnO nanocables through thermal oxidation of Zn nanowires grown by RF magnetron sputtering. *J. Cryst. Growth* 2006, 290, 485-489.

[70] Choopun, S.; Hongsith, N.; Tanunchai, S.; Chairuangsri, T.; Krua-in, C.; Singkarat, S.; Vialithong, T.; Mangkorntong, P.; Mangkorntong, N., Single-crystalline ZnO nanobelts by RF sputtering. *J. Cryst. Growth* 2005, 282, 365-369.

[71] Chen, M.-T.; Ting, J.-M., Sputter deposition of ZnO nanorods/thin-film structures on Si. *Thin Solid Films* 2006, 494, 250-254.

[72] Usuda, M.; Hamada, N.; Kotani, T.; van Schilfgaarde, M., All-electron GW calculation based on the LAPW method: Application to wurtzite ZnO. *Phys. Rev. B* 2002, 66 (12).

[73] Jaffe, J. E.; Snyder, J. A.; Lin, Z. J.; Hess, A. C., LDA and GGA calculations for high-pressure phase transitions in ZnO and MgO. *Phys. Rev. B* 2000, 62 (3), 1660-1665.

[74] Chelikowsky, J. R., Oxygen Pseudopotential - Application to Electronic-Structure of Zno. *Solid State Commun.* 1977, 22 (6), 351-354.

[75] Rössler, U., Energy Bands of Hexagonal II-VI Semiconductors. *Phys. Rev.* 1969, 184, 733–738.

[76] Bloom, S.; Ortenbur.I, Pseudopotential Band-Structure of Zno. *Phys. Status Solidi B* 1973, 58 (2), 561-566.

[77] Ivanov, I.; Pollmann, J., Electronic-Structure of Ideal and Relaxed Surfaces of Zno - a Prototype Ionic Wurtzite Semiconductor and Its Surface-Properties. *Phys. Rev. B* 1981, 24 (12), 7275-7296.

[78] Vogel, D.; Kruger, P.; Pollmann, J., Ab-Initio Electronic-Structure Calculations for Ii-Vi Semiconductors Using Self-Interaction-Corrected Pseudopotentials. *Phys. Rev. B* 1995, 52 (20), 14316-14319.

[79] Broser, I.; Broser, R.; Finkenrath, H.; Galazka, R. R.; Gumlich, H. E.; Hoffmann, A.; Kossut, J.; Mollwo, E.; Nelkowski, H.; Nimtz, G.; Osten, W. v. d.; Rosenzweig, M.; Schulz, H. J.; Theis, D.; Tschierse, D., Physics of II-

VI and I-VII Compounds, Semimagnetic Semiconductors. Springer: 1982; Vol. 17b, p 543.

[80] Sheng, H.; Muthukumar, S.; Emanetoglu, N. W.; Lu, Y., Schottky diode with Ag on (11(2)over-bar0) epitaxial ZnO film. *Appl. Phys. Lett.* 2002, 80 (12), 2132-2134.

[81] Minami, T.; Sato, H.; Nanto, H.; Takata, S., Group III Impurity Doped Zinc Oxide Thin Films Prepared by RF Magnetron Sputtering. *J. Appl. Phys.* 1985, 24, L781-L784.

[82] Zhang, S. B.; Wei, S. H.; Zunger, A., Intrinsic n-type versus p-type doping asymmetry and the defect physics of ZnO. *Phys. Rev. B* 2001, 63 (7), 205.

[83] Lee, E. C.; Kim, Y. S.; Jin, Y. G.; Chang, K. J., Compensation mechanism for N acceptors in ZnO. *Phys. Rev. B* 2001, 64 (8), 085120.

[84] Van de Walle, C. G., Hydrogen as a cause of doping in zinc oxide. *Phys. Rev. Lett.* 2000, 85 (5), 1012-1015.

[85] Albrecht, J. D.; Ruden, P. P.; Limpijumnong, S.; Lambrecht, W. R. L.; Brennan, K. F., High field electron transport properties of bulk ZnO. *J. Appl. Phys.* 1999, 86 (12), 6864-6867.

[86] Look, D. C., Recent advances in ZnO materials and devices. *Mat. Sci. Eng. B-Solid* 2001, 80 (1-3), 383-387.

[87] Mead, C. A., Metal-semiconductor surface barriers. *Solid-State Electron.* 1966, 9 (11-12), 1023-1033.

[88] Allen, M. W.; Alkaisi, M. M.; Durbin, S. M., Metal Schottky diodes on Zn-polar and O-polar bulk ZnO. *Appl. Phys. Lett.* 2006, 89 (10), 103520.

[89] Allen, M. W.; Durbin, S. M., Influence of oxygen vacancies on Schottky contacts to ZnO. *Appl. Phys. Lett.* 2008, 92 (12), 053512.

[90] Allen, M. W.; Durbin, S. M.; Metson, J. B., Silver oxide Schottky contacts on n-type ZnO. *Appl. Phys. Lett.* 2007, 91 (5), 122110.

[91] Brillson, L. J.; Mosbacker, H. L.; Hetzer, M. J.; Strzhemechny, Y.; Jessen, G. H.; Look, D. C.; Cantwell, G.; Zhang, J.; Song, J. J., Dominant effect of near-interface native point defects on ZnO Schottky barriers. *Appl. Phys. Lett.* 2007, 90 (10), 102116.

[92] Mosbacker, H. L.; El Hage, S.; Gonzalez, M.; Ringel, S. A.; Hetzer, M.; Look, D. C.; Cantwell, G.; Zhang, J.; Song, J. J.; Brillson, L. J., Role of subsurface defects in metal-ZnO(0001) Schottky barrier formation. *J. Vac. Sci. Technol. B* 2007, 25 (4), 1405-1411.

[93] Nakano, M.; Tsukazaki, A.; Gunji, R. Y.; Ueno, K.; Ohtomo, A.; Fukumura, T.; Kawasaki, M., Schottky contact on a ZnO (0001) single crystal with conducting polymer. *Appl. Phys. Lett.* 2007, 91 (14), 142113.

[94] Oh, M. S.; Hwang, D. K.; Lim, J. H.; Choi, Y. S.; Park, S. J., Improvement of Pt Schottky contacts to n-type ZnO by KrF excimer laser irradiation. *Appl. Phys. Lett.* 2007, 91 (4), 042109.

[95] Gunji, R. Y.; Nakano, M.; Tsukazaki, A.; Ohtomo, A.; Fukumura, T.; Kawasaki, M., Polymer Schottky contact on O-polar ZnO with silane coupling agent as surface protective layer. *Appl. Phys. Lett.* 2008, 93 (1), 012104.

[96] S. Kurtin, T. C. M., and C. A. Mead, Fundamental Transition In The Electronic Nature Of Solids. *Phys. Rev. Lett.* 1969, 22, 1433–1436.

[97] Schottky, W., Halbleitertheorie der Sperrschicht. *Die Naturwissenschaften* 1938, 26 (52), 843-843.

[98] Polyakov, A. Y.; Smirnov, N. B.; Kozhukhova, E. A.; Vdovin, V. I.; Ip, K.; Norton, D. P.; Pearton, S. J., Properties of Au and Ag Schottky diodes prepared on undoped n-ZnO. *J. Vac. Sci. Technol. B* 2003, 21 (5), 1603-1608.

[99] Polyakov, A. Y.; Smirnov, N. B.; Kozhukhova, E. A.; Vdovin, V. I.; Ip, K.; Heo, Y. W.; Norton, D. P.; Pearton, S. J., Electrical characteristics of Au and Ag Schottky contacts on n-ZnO. *Appl. Phys. Lett.* 2003, 83 (8), 1575-1577.

[100] Oh, D. C.; Kim, J. J.; Makino, H.; Hanada, T.; Cho, M. W.; Yao, T.; Ko, H. J., Characteristics of Schottky contacts to ZnO : N layers grown by molecular-beam epitaxy. *Appl. Phys. Lett.* 2005, 86 (4), 042110.

[101] Kim, S. H.; Kim, H. K.; Seong, T. Y., Effect of hydrogen peroxide treatment on the characteristics of Pt Schottky contact on n-type ZnO. *Applied Physics Letters* 2005, 86 (11), 112101.

[102] Kim, S.; Kang, B. S.; Ren, F.; Ip, K.; Heo, Y. W.; Norton, D. P.; Pearton, S. J., Sensitivity of Pt/ZnO Schottky diode characteristics to hydrogen. *Appl. Phys. Lett.* 2004, 84 (10), 1698-1700.

[103] Ip, K.; Heo, Y. W.; Baik, K. H.; Norton, D. P.; Pearton, S. J.; Ren, F., Specific contact resistance of Ti/Al/PVAu ohmic contacts to photphorus-doped ZnO thin films. *J. Vac. Sci. Technol. B* 2004, 22 (1), 171-174.

[104] Kim, S. H.; Kim, H. K.; Seong, T. Y., Electrical characteristics of Pt Schottky contacts on sulfide-treated n-type ZnO. *Appl. Phys. Lett.* 2005, 86 (2), 022101.

[105] Ip, K.; Gila, B. P.; Onstine, A. H.; Lambers, E. S.; Heo, Y. W.; Baik, K. H.; Norton, D. P.; Pearton, S. J.; Kim, S.; LaRoche, J. R.; Ren, F., Improved Pt/Au and W/Pt/Au Schottky contacts on n-type ZnO using ozone cleaning. *Appl. Phys. Lett.* 2004, 84 (25), 5133-5135.

[106] Ip, K.; Heo, Y. W.; Baik, K. H.; Norton, D. P.; Pearton, S. J.; Ren, F., Carrier concentration dependence of Ti/Al/Pt/Au contact resistance on n-type ZnO. *Appl. Phys. Lett.* 2004, 84 (4), 544-546.

[107] von Wenckstern, H.; Biehne, G.; Rahman, R. A.; Hochmuth, H.; Lorenz, M.; Grundmann, M., Mean barrier height of Pd Schottky contacts on ZnO thin films. *Appl. Phys. Lett.* 2006, 88 (9), 092192.

[108] Grossner, U.; Gabrielsen, S.; Borseth, T. M.; Grillenberger, J.; Kuznetsov, A. Y.; Svensson, B. G., Palladium Schottky barrier contacts to hydrothermally grown n-ZnO and shallow electron states. *Appl. Phys. Lett.* 2004, 85 (12), 2259-2261.

[109] Schifano, R.; Monakhov, E. V.; Grossner, U.; Svensson, B. G., Electrical characteristics of palladium Schottky contacts to hydrogen peroxide treated hydrothermally grown ZnO. *Appl. Phys. Lett.* 2007, 91 (19), 193507.

[110] Janotti, A.; Van de Walle, C. G., Oxygen vacancies in ZnO. *Appl. Phys. Lett.* 2005, 87 (12), 122102.

[111] Lao, C. S.; Liu, J.; Gao, P. X.; Zhang, L. Y.; Davidovic, D.; Tummala, R.; Wang, Z. L., ZnO nanobelt/nanowire Schottky diodes formed by dielectrophoresis alignment across Au electrodes. *Nano Letters* 2006, 6 (2), 263-266.

[112] Kim, H. K.; Han, S. H.; Seong, T. Y.; Choi, W. K., Low-resistance Ti/Au ohmic contacts to Al-doped ZnO layers. *Applied Physics Letters* 2000, 77 (11), 1647-1649.

[113] Lee, J. M.; Kim, K. K.; Park, S. J.; Choi, W. K., Low-resistance and nonalloyed ohmic contacts to plasma treated ZnO. *Appl. Phys. Lett.* 2001, 78 (24), 3842-3844.

[114] Kim, H. K.; Kim, K. K.; Park, S. J.; Seong, T. Y.; Adesida, I., Formation of low resistance nonalloyed Al/Pt ohmic contacts on n-type ZnO epitaxial layer. *J. Appl. Phys.* 2003, 94 (6), 4225-4227.

[115] Kim, H. K.; Adesida, I.; Kim, K. K.; Park, S. J.; Seong, T. Y., Study of the electrical and structural characteristics of Al/Pt ohmic contacts on n-type ZnO epitaxial layer. *J. Electrochem. Soc.* 2004, 151 (4), G223-G226.

[116] Lim, J. H.; Kim, K. K.; Hwang, D. K.; Kim, H. S.; Oh, J. Y.; Park, S. J., Formation and effect of thermal annealing for low-resistance Ni/Au ohmic

contact to phosphorous-doped p-type ZnO. *J. Electrochem. Soc.* 2005, 152 (3), G179-G181.

[117] Kang, S. H.; Hwang, D. K.; Park, S. J., Low-resistance and highly transparent Ni/indium-tin oxide ohmic contacts to phosphorous-doped p-type ZnO. *Appl. Phys. Lett.* 2005, 86 (21), 211902.

[118] Mandalapu, L. J.; Yang, Z.; Liu, J. L., Low-resistivity Au/Ni ohmic contacts to sb-doped p-type ZnO. *Applied Physics Letters* 2007, 90 (25).

[119] Kim, S. H.; Kim, K. K.; Park, S. J.; Seong, T. Y., Thermally stable and low resistance Re/Ti/Au ohmic contacts to n-ZnO. *J. Electrochem. Soc.* 2005, 152 (3), G169-G172.

[120] Qin, Y.; Wang, X. D.; Wang, Z. L., Microfibre-nanowire hybrid structure for energy scavenging. *Nature* 2008, 451 (7180), 809-813.

[121] Hochbaum, A. I.; Chen, R. K.; Delgado, R. D.; Liang, W. J.; Garnett, E. C.; Najarian, M.; Majumdar, A.; Yang, P. D., Enhanced thermoelectric performance of rough silicon nanowires. *Nature* 2008, 451 (7175), 163-167.

[122] Chan, C. K.; Peng, H. L.; Liu, G.; McIlwrath, K.; Zhang, X. F.; Huggins, R. A.; Cui, Y., High-performance lithium battery anodes using silicon nanowires. *Nat. Nanotechnol.* 2008, 3 (1), 31-35.

[123] Guo, J.; Wang, J.; Polizzi, E.; Datta, S.; Lundstrom, M., Electrostatics of nanowire transistors. *Ieee T. Nanotechnol.* 2003, 2 (4), 329-334.

[124] Yeo, Y. C., Metal gate technology for nanoscale transistors - material selection and process integration issues. *Thin Solid Films* 2004, 462, 34-41.

[125] LaRoche, J. R.; Heo, Y. W.; Kang, B. S.; Tien, L. C.; Kwon, Y.; Norton, D. P.; Gila, B. P.; Ren, F.; Pearton, S. J., Fabrication approaches to ZnO nanowire devices. *J. Electron. Mater.* 2005, 34 (4), 404-408.

[126] Heo, Y. W.; Norton, D. P.; Tien, L. C.; Kwon, Y.; Kang, B. S.; Ren, F.; Pearton, S. J.; LaRoche, J. R., ZnO nanowire growth and devices. *Mat. Sci. Eng. R* 2004, 47 (1-2), 1-47.

[127] Pearton, S., ZnO Nanowire Devices. *The Journal of The Minerals, Metals & Materials Society* 2004, 56, 247-248.

[128] Lin, Y. M.; Appenzeller, J.; Knoch, J.; Avouris, P., High-performance carbon nanotube field-effect transistor with tunable Polarities. *Ieee T. Nanotechnol.* 2005, 4 (5), 481-489.

[129] Huang, Y.; Duan, X. F.; Cui, Y.; Lauhon, L. J.; Kim, K. H.; Lieber, C. M., Logic gates and computation from assembled nanowire building blocks. *Science* 2001, 294 (5545), 1313-1317.

[130] Heo, Y. W.; Tien, L. C.; Kwon, Y.; Norton, D. P.; Pearton, S. J.; Kang, B. S.; Ren, F., Depletion-mode ZnO nanowire field-effect transistor. *Appl. Phys. Lett.* 2004, 85 (12), 2274-2276.

[131] Keem, K.; Kang, J.; Yoon, C.; Yeom, D.; Jeong, D. Y.; Moon, B. M.; Kim, S., A fabrication technique for top-gate ZnO nanowire field-effect transistors by a photolithography process. *Microelectron. Eng.* 2007, 84 (5-8), 1622-1626.

[132] Keem, K.; Jeong, D. Y.; Kim, S.; Lee, M. S.; Yeo, I. S.; Chung, U. I.; Moon, J. T., Fabrication and device characterization of omega-shaped-gate ZnO nanowire field-effect transistors. *Nano Lett.* 2006, 6 (7), 1454-1458.

[133] Ng, H. T.; Han, J.; Yamada, T.; Nguyen, P.; Chen, Y. P.; Meyyappan, M., Single crystal nanowire vertical surround-gate field-effect transistor. *Nano Lett.* 2004, 4 (7), 1247-1252.

[134] Cha, S. N.; Jang, J. E.; Choi, Y.; Amaratunga, G. A. J.; Ho, G. W.; Welland, M. E.; Hasko, D. G.; Kang, D. J.; Kim, J. M., High performance ZnO nanowire field effect transistor using self-aligned nanogap gate electrodes. *Appl. Phys. Lett.* 2006, 89 (26), 263102.

[135] Wade, T. L.; Hoffer, X.; Mohammed, A. D.; Dayen, J. F.; Pribat, D.; Wegrowe, J. E., Nanoporous alumina wire templates for surrounding-gate nanowire transistors. *Nanotechnology* 2007, 18 (12), 125201.

[136] Park, W. I.; Kim, J. S.; Yi, G. C.; Bae, M. H.; Lee, H. J., Fabrication and electrical characteristics of high-performance ZnO nanorod field-effect transistors. *Appl. Phys. Lett.* 2004, 85 (21), 5052-5054.

[137] Chang, P. C.; Fan, Z. Y.; Wang, D. W.; Tseng, W. Y.; Chiou, W. A.; Hong, J.; Lu, J. G., ZnO nanowires synthesized by vapor trapping CVD method. *Chem. Mater.* 2004, 16 (24), 5133-5137.

[138] Fan, Z. Y.; Wang, D. W.; Chang, P. C.; Tseng, W. Y.; Lu, J. G., ZnO nanowire field-effect transistor and oxygen sensing property. *Appl. Phys. Lett.* 2004, 85 (24), 5923-5925.

[139] Hossain, F. M.; Nishii, J.; Takagi, S.; Sugihara, T.; Ohtomo, A.; Fukumura, T.; Koinuma, H.; Ohno, H.; Kawasaki, M., Modeling of grain boundary barrier modulation in ZnO invisible thin film transistors. *Physica E* 2004, 21 (2-4), 911-915.

[140] Fortunato, E. M. C.; Barquinha, P. M. C.; Pimentel, A. C. M. B. G.; Goncalves, A. M. F.; Marques, A. J. S.; Martins, R. F. P.; Pereira, L. M. N., Wide-bandgap high-mobility ZnO thin-film transistors produced at room temperature. *Appl. Phys. Lett.* 2004, 85 (13), 2541-2543.

[141] Xu, C. X.; Sun, X. W.; Chen, B. J., Field emission from gallium-doped zinc oxide nanofiber array. *Appl. Phys. Lett.* 2004, 84 (9), 1540-1542.

[142] Jie, J. S.; Wang, G. Z.; Han, X. H.; Yu, Q. X.; Liao, Y.; Li, G. P.; Hou, J. G., Indium-doped zinc oxide nanobelts. *Chem. Phys. Lett.* 2004, 387 (4-6), 466-470.

[143] Ronning, C.; Gao, P. X.; Ding, Y.; Wang, Z. L.; Schwen, D., Manganese-doped ZnO nanobelts for spintronics. *Appl. Phys. Lett.* 2004, 84 (5), 783-785.

[144] Geng, B. Y.; Wang, G. Z.; Jiang, Z.; Xie, T.; Sun, S. H.; Meng, G. W.; Zhang, L. D., Synthesis and optical properties of S-doped ZnO nanowires. *Appl. Phys. Lett.* 2003, 82 (26), 4791-4793.

[145] Bae, S. Y.; Seo, H. W.; Park, J. H., Vertically aligned sulfur-doped ZnO nanowires synthesized via chemical vapor deposition. *J. Phys. Chem. B* 2004, 108 (17), 5206-5210.

[146] He, J. H.; Lao, C. S.; Chen, L. J.; Davidovic, D.; Wang, Z. L., Large-scale Ni-doped ZnO nanowire arrays and electrical and optical properties. *J. Am. Chem. Soc.* 2005, 127 (47), 16376-16377.

[147] Lin, D. D.; Wu, H.; Pan, W., Photoswitches and Memories Assembled by Electrospinning Aluminum-Doped Zinc Oxide Single Nanowires. *Adv. Mater.* 2007, 19 (22), 3968-3972.

[148] Chen, P. C.; Sukcharoenchoke, S.; Ryu, K.; de Arco, L. G.; Badmaev, A.; Wang, C.; Zhou, C. W., 2,4,6-Trinitrotoluene (TNT) Chemical Sensing Based on Aligned Single-Walled Carbon Nanotubes and ZnO Nanowires. *Adv. Mater.* 2010, 22 (17), 1900-1902.

[149] Al-Hilli, S. M.; Willander, M.; Ost, A.; Stralfors, P., ZnO nanorods as an intracellular sensor for pH measurements. *J. Appl. Phys.* 2007, 102 (8), 014302.

[150] Al-Hilli, S. M.; Al-Mofarji, R. T.; Klason, P.; Willander, M.; Gutman, N.; Sa'ar, A., Zinc oxide nanorods grown on two-dimensional macroporous periodic structures and plane Si as a pH sensor. *J. Appl. Phys.* 2008, 103 (1), 084304.

[151] Motto, P.; Cauda, V.; Stassi, S.; Canavese, G.; Demarchi, D., Functionalized single ZnO-metal junction as a pH sensor. In IEEE Sensors 2013, Baltimora, 2013; Vol. doi:10.1109/ICSENS.2013. 6688513, pp 1-4.

[152] Youn, S. K.; Ramgir, N.; Wang, C. Y.; Subannajui, K.; Cimalla, V.; Zacharias, M., Catalyst-Free Growth of ZnO Nanowires Based on Topographical Confinement and Preferential Chemisorption and Their Use

for Room Temperature CO Detection. *J. Phys. Chem. C* 2010, 114 (22), 10092-10100.

[153] Calestani, D.; Zha, M.; Mosca, R.; Zappettini, A.; Carotta, M. C.; Di Natale, V.; Zanotti, L., Growth of ZnO tetrapods for nanostructure-based gas sensors. *Sens. Actuators, B* 2010, 144 (2), 472-478.

[154] Zhu, R.; Wang, D. Q.; Xiang, S. Q.; Zhou, Z. Y.; Ye, X. Y., Piezoelectric characterization of a single zinc oxide nanowire using a nanoelectromechanical oscillator. *Nanotechnology* 2008, 19 (28).

[155] Zhu, R.; Wang, D. Q.; Xiang, S. Q.; Zhou, Z. Y.; Ye, X. Y., Zinc oxide nanowire electromechanical oscillator. *Sensor Actuat a-Phys.* 2009, 154 (2), 224-228.

[156] Soci, C.; Zhang, A.; Xiang, B.; Dayeh, S. A.; Aplin, D. P. R.; Park, J.; Bao, X. Y.; Lo, Y. H.; Wang, D., ZnO nanowire UV photodetectors with high internal gain. *Nano Lett.* 2007, 7 (4), 1003-1009.

[157] Suehiro, J.; Nakagawa, N.; Hidaka, S.; Ueda, M.; Imasaka, K.; Higashihata, M.; Okada, T.; Hara, M., Dielectrophoretic fabrication and characterization of a ZnO nanowire-based UV photosensor. *Nanotechnology* 2006, 17 (10), 2567-2573.

[158] Das, S. N.; Kar, J. P.; Choi, J. H.; Il Lee, T.; Moon, K. J.; Myoung, J. M., Fabrication and Characterization of ZnO Single Nanowire-Based Hydrogen Sensor. *J. Phys. Chem.* C 2010, 114 (3), 1689-1693.

[159] Wang, J. X.; Sun, X. W.; Yang, Y.; Huang, H.; Lee, Y. C.; Tan, O. K.; Vayssieres, L., Hydrothermally grown oriented ZnO nanorod arrays for gas sensing applications. *Nanotechnology* 2006, 17, 4995.

[160] Qi, Q.; Zhang, T.; Liu, L.; Zheng, X.; Yu, Q.; Zeng, Y.; H. Yang, Selective acetone sensor based on dumbbell-like ZnO with rapid response and recovery. *Sens. Actuators, B* 2008, 134, 166.

[161] Wang, D.; Chu, X.; Gong, M., Hydrothermal growth of ZnO nanoscrew drivers and their gas sensing properties. *Nanotechnology* 2007, 18, 185601.

[162] Kakati, N.; Jee, S. H.; Kim, S. H.; Lee, H. K.; Yoon, Y. S., Sensitivity enhancement of ZnO nanorod gas sensors with surface modification by InSb thin film. *Jpn. J. Appl. Phys.* 2009, 48, 105002.

[163] Ogata, K.; Koike, K.; Sasa, S.; Inoue, M.; Yano, M., ZnO nanorod arrays on n-type Si(111) substrates for pH measurements. *J. Vac. Sci. Technol., B* 2009, 27, 1684.

[164] Yales, D. E.; Levine, S.; Healy, T. W., Site-binding model of the electrical double layer at the oxide/water interface *J. Chem. Soc., Faraday Trans.* 1974, 70, 1807-1818.

[165] Siu, W. M.; Collold, R. S. C., Basic properties of the electrolyte—SiO2—Si system: Physical and theoretical aspects. *IEEE Trans. Electron. Devices* 1979, 26, 1805-1815.

[166] Kang, B. S.; Ren, F.; Heo, Y. W.; Tien, L. C.; Norton, D. P.; Pearton, S. J., pH measurements with single ZnO nanorods integrated with a microchannel *Appl. Phys. Lett.* 2005, 86, 112105.

[167] Cauda, V.; Motto, P.; Perrone, D.; Demarchi, D.; Piccinini, G., pH-triggered conduction of amine-functionalized single ZnO wire integrated on a customized nanogap electronic platform. *Nanoscale Res. Lett.* 2014,9, 1-10.

[168] Soci, C.; Zhang, A.; Xiang, B.; Dayeh, S. A.; Aplin, D. P. R.; Park, J.; Bao, X. Y.; Lo, Y. H.; Wang, D., ZnO Nanowire UV Photodetectors with High Internal Gain. *Nano Lett.* 2007, 7, 1003-1009.

[169] Yang, Q.; Guo, X.; Wang, W.; Zhang, Y.; Xu, S.; Lien, D. H.; Wang, Z. L., Enhancing Sensitivity of a Single ZnO Micro-/Nanowire Photodetector by Piezo-phototronic Effect. *ACS Nano* 2010, 4 (10), 6285-6291.

[170] Wang, Z. L., Triboelectric nanogenerators as new energy technology for self-powered systems and as active mechanical and chemical sensors. *ACS Nano* 2013, 7 (11), 9533-9557.

[171] Wang, X.; Song, J.; Liu, J.; Zhong, L. W., Direct-current nanogenerator driven by ultrasonic waves. *Science* 2007, 316 (5821), 102-105.

[172] Xu, S.; Wei, Y.; Liu, J.; Yang, R.; Wang, Z. L., Integrated multilayer nanogenerator fabricated using paired nanotip-to-nanowire brushes. *Nano Lett.* 2008, 8 (11), 4027-4032.

[173] Wang, Z. L.; Yang, R. S.; Zhou, J.; Qin, Y.; Xu, C.; Hu, Y. F.; Xu, S., Lateral nanowire/nanobelt based nanogenerators, piezotronics and piezo-phototronics. *Materials Science & Engineering R-Reports* 2010, 70 (3-6), 320-329.

[174] Qin, Y.; Wang, X.; Wang, Z. L., Microfibre-nanowire hybrid structure for energy scavenging. *Nature* 2008, 451 (7180), 809-813.

[175] Lee, M.; Chen, C. Y.; Wang, S.; Cha, S. N.; Park, Y. J.; Kim, J. M.; Chou, L. J.; Wang, Z. L., A hybrid piezoelectric structure for wearable nanogenerators. *Adv. Mater.* 2012, 24 (13), 1759-1764.

[176] Hu, Y.; Lin, L.; Zhang, Y.; Wang, Z. L., Replacing a battery by a nanogenerator with 20 v output. *Adv. Mater.* 2012, 24 (1), 110-114.

[177] Zhu, G.; Wang, A. C.; Liu, Y.; Zhou, Y.; Wang, Z. L., Functional electrical stimulation by nanogenerator with 58 v output voltage. *Nano Lett.* 2012, 12 (6), 3086-3090.

[178] Fan, F. R.; Tian, Z. Q.; Lin Wang, Z., Flexible triboelectric generator. *Nano Energy* 2012, 1 (2), 328-334.

[179] Zhu, G.; Lin, Z. H.; Jing, Q.; Bai, P.; Pan, C.; Yang, Y.; Zhou, Y.; Wang, Z. L., Toward large-scale energy harvesting by a nanoparticle-enhanced triboelectric nanogenerator. *Nano Lett.* 2013, 13 (2), 847-853.

[180] Zhang, S.; Yang, X.; Numata, Y.; Han, L., Highly efficient dye-sensitized solar cells: progress and future challenges. *Energy & Environmental Science* 2013, 6 (5), 1443-1464.

[181] Youngblood, W. J.; Lee, S.-H. A.; Maeda, K.; Mallouk, T. E., Visible light water splitting using dye-sensitized oxide semiconductors. *Accounts of chemical research* 2009, 42 (12), 1966-1973.

[182] Lupan, O.; Guérin, V.; Tiginyanu, I.; Ursaki, V.; Chow, L.; Heinrich, H.; Pauporté, T., Well-aligned arrays of vertically oriented ZnO nanowires electrodeposited on ITO-coated glass and their integration in dye sensitized solar cells. *Journal of Photochemistry and Photobiology A: Chemistry* 2010, 211 (1), 65-73.

[183] Hsu, C.-H.; Chen, D.-H., Photoresponse and stability improvement of ZnO nanorod array thin film as a single layer of photoelectrode for photoelectrochemical water splitting. *International Journal of Hydrogen Energy* 2011, 36 (24), 15538-15547.

[184] Maçaira, J.; Andrade, L.; Mendes, A., Review on nanostructured photoelectrodes for next generation dye-sensitized solar cells. *Renewable and Sustainable Energy Reviews* 2013, 27, 334-349.

[185] Omar, A.; Abdullah, H., Electron transport analysis in zinc oxide-based dye-sensitized solar cells: A review. *Renewable and Sustainable Energy Reviews* 2014, 31, 149-157.

[186] Pérez-Larios, A.; Lopez, R.; Hernández-Gordillo, A.; Tzompantzi, F.; Gómez, R.; Torres-Guerra, L., Improved hydrogen production from water splitting using TiO< sub> 2</sub>–ZnO mixed oxides photocatalysts. *Fuel* 2012, 100, 139-143.

[187] Yang, X.; Wolcott, A.; Wang, G.; Sobo, A.; Fitzmorris, R. C.; Qian, F.; Zhang, J. Z.; Li, Y., Nitrogen-doped ZnO nanowire arrays for photoelectrochemical water splitting. *Nano letters* 2009, 9 (6), 2331-2336.

[188] Yun, S.; Lim, S., Improved conversion efficiency in dye-sensitized solar cells based on electrospun Al-doped ZnO nanofiber electrodes prepared by seed layer treatment. *Journal of Solid State Chemistry* 2011, 184 (2), 273-279.

[189] Yu, D.; Trad, T.; McLeskey Jr, J. T.; Craciun, V.; Taylor, C. R., ZnO nanowires synthesized by vapor phase transport deposition on transparent oxide substrates. *Nanoscale research letters* 2010, 5 (8), 1333-1339.

In: Nanowires
Editor: Luke J. Wilson, pp. 59-87
ISBN: 978-1-63117-855-9

Chapter 2

ROUGH SILVER NANOWIRE, NANOBUD AND NANOPARTICLE SUBSTRATES: PREPARATION, PROPERTIES AND USE IN THE SERS DETECTION OF BIOMACROMOLECULES

Shu Zhang[1], Wanyun Ma[1,2], Qun Wang[3], Dapeng Xu[1], Chuyun Deng[1] and Jia-Lin Sun[*,1,2]

[1] State Key Lab of Low-Dimensional Quantum Physics, Department of Physics, Tsinghua University, Beijing, China
[2] Collaborative Innovation Center of Quantum Matter, Beijing, China
[3] College of Information, Beijing City University, Beijing, China

Abstract

This chapter introduces an improved solid-state ionics method combined with vacuum thermal evaporation to synthesize silver nanowire, nanobud and nanoparticle structures. Surface enhancement effects in the silver nanostructures are studied, and provide a reliable approach towards the trace analysis of biomacromolecules. First, we introduce the solid-state ionics method for synthesizing silver nanostructures, and discuss morphological changes resulting from different growth currents. Second, we discuss surface plasmon polaritons

[*] E-mail address: jlsun@tsinghua.edu.cn (Corresponding author)

excited on the silver nanowires. Third, we detect the surface-enhanced Raman scattering signal of low concentrations of Rhodamine 6G on silver nanostructure substrates. This demonstrates that the rough silver nanostructure surface provides a significant surface enhancement. Fourth, the effects of natural oxidation on the surface enhancement of the silver nanostructures are studied. The effect of oxidation on the silver nanostructure substrates, such as their thickness, surface roughness and particles size, is discussed. Oxidation strongly affects the surface enhancement of the silver nanostructures. Finally, the nanostructures are applied in the surface-enhanced detection of λ-DNA. The Raman signal of a single λ-DNA molecule is detected, as a result of the surface enhancement by the silver nanostructured substrate. The enhanced fluorescent signal of λ-DNA is detected at more than 10 μm from the substrate surface. The rough silver nanostructures exhibit excellent surface-enhanced Raman scattering, and provide a new approach for biomedical research.

1. Introduction

Many theoretical and experimental studies have shown that surface plasmon resonance (SPR) readily occurs on noble metal nanostructure surfaces, when excited by incident photons. This results in a significant local field enhancement, leading to surface-enhanced Raman scattering (SERS) and surface-enhanced fluorescence (SEF). SERS is a powerful tool in chemical analysis for studying molecular vibrations. The Fleishmann group first observed the SERS phenomenon on a silver electrode in 1974 [1]. SERS can enhance the Raman scattering intensity of molecules adsorbed on the surface, by more than 10 orders of magnitude. The sensitivity of SERS spectroscopy can therefore reach single molecule detection levels [2-8]. SEF is the significant increase in emission intensity of fluorescent molecules near the metal surface, compared with their fluorescence in the free state [9-12]. The Lakowicz group, Maryland University, USA, has been studying SEF since 1999, and has reported numerous important achievements [11-12]. They demonstrated that the fluorescence signal could be enhanced by increasing the emission attenuation rate. The preparation of efficient enhancement substrates and the study of the enhancement mechanism have become core issues, regarding how to improve the surface enhancement effect [13-17]. The most suitable substrates for surface enhancement to date have been noble metal nanostructures.

Sliver is often used in surface enhancement substrates, since gold is significantly more expensive. In this chapter, an improved solid-state ionics method for synthesizing large-area silver nanowires (AgNWs) and silver nanobuds (AgNBs) is introduced. By controlling the impressed current (*I*), we obtained different silver nanostructures with different morphologies and regular arrangements. We carried out a detailed detection of the surface enhancement of these substrates, with the molecular probe Rhodamine 6G (R6G).

The surface oxidation of silver nanostructures is inherently problematic, since silver is easily oxidized in air. This is especially so for rough silver nanostructure surfaces, with large surface area to volume ratios [18-22]. For this reason, we explore the effect of oxidation of silver nanoparticle (AgNP) films on the surface enhancement effect. AgNP films were prepared by vacuum thermal evaporation. Details of how oxidation affected the surface enhancement effect of these silver nanostructure substrates, such as their change in thickness, surface roughness and particles size, are discussed.

Last, AgNBs were used to detect the Raman and fluorescence signals of λ-DNA. The Raman signal was significantly enhanced on the AgNB surface, and reached single molecule detection limits. This revealed information about λ-DNA as a whole, and even information about specific parts of a single λ-DNA molecule. The fluorescence signal of λ-DNA was detected at more than 10 μm from the edges of the AgNB, which indicated that fluorescence enhancement occurred over a relatively long distance.

2. Preparation of Rough AgNW, AgNB and AgNP Substrates

2.1. Preparation of Silver Nanostructures by Solid-State Ionics Method

An improved solid-state ionics method was used to synthesize large-area AgNW and AgNB substrates, using superionic conducting $RbAg_4I_5$ films under a direct current electric field (DCEF). A measuring Source-Meter (Keithley 2400) was used to control the *I*.

Figure 1 shows a schematic diagram of the apparatus used to synthesis the silver nanostructures. The 400-nm-thick superionic conductor $RbAg_4I_5$ film was deposited by vacuum thermal evaporation on a clean glass substrate [23, 24]. Two parallel 1-μm-thick silver films separated by ~9 cm were deposited by vacuum thermal evaporation on the ends of the $RbAg_4I_5$ film, for use as electrodes. Evaporation was carried out at a pressure of $\sim 1\times10^{-4}$ Pa at room temperature.

Ag ions were not bound within a fixed crystal lattice in the $RbAg_4I_5$ film, but travelled in "ion channels" under the DCEF [23, 24]. Silver atoms initially lost electrons to become silver ions on the anode surface, and then transferred to the cathode through "ion channels" in the $RbAg_4I_5$ film. Electrons were then transferred to the cathode, by a conducting wire in the external circuit. Silver ions gained electrons and were reduced to silver atoms, at the edge of the cathode. Silver nanostructures were obtained by the accumulation and crystallization of silver atoms at the cathode. Thus, the silver film anode served as both an electrode and a source of silver ions, while the $RbAg_4I_5$ film served as a medium for transmitting silver ions.

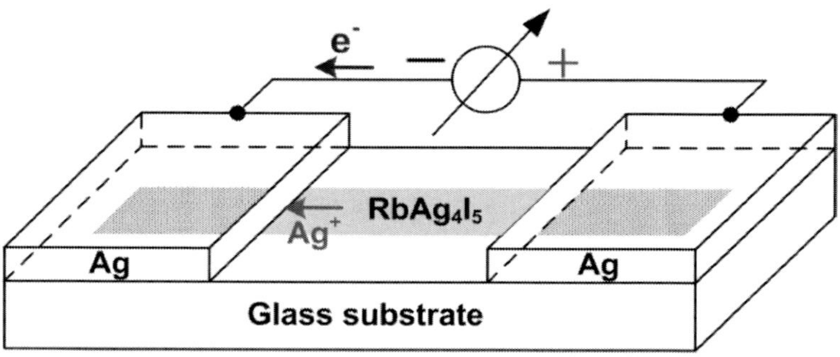

Figure 1. Schematic diagram of the apparatus used to synthesize the silver nanostructures.

Figure 2. Photograph of an aligned silver nanostructure at the cathode, and depleted silver film at the anode.

We refer to this as the improved solid-state ionics method. It is related to other techniques used to prepare silver nanostructures, but has many advantages. First, additional steps such as separating Ag nanostructures from solutions were not required, since the experiment was carried out in the solid phase. Avoiding these steps helped retain the Ag nanostructures in their original state. Second, the direction of the nanostructures could be controlled by the electric field, so highly aligned nanostructures could be produced. Third, the constant direct current conditions resulted in the constant accumulation of silver atoms, so very large-area silver nanostructures could be synthesized. The area of the silver nanostructure reached ~1 cm^2, after more than 50 hours' growth time (Figure 2).

2.2. Morphology and Composition of Silver Nanostructures Prepared by the Solid-State Ionics Method under Different Applied DCEFs [25]

Silver nanostructures with different morphologies were prepared by varying the *I* between the two silver film electrodes. Figure 3a, b, c and d show the microscopic surface morphologies of silver nanostructures prepared at *I* of 3, 5, 8 and 12 μA, respectively. The same vacuum evaporation conditions were used as those in Figure 1.

Figure 3 shows that the morphologies of the silver nanostructures differed when prepared at different *I*. Disordered AgNWs were obtained when prepared at an *I* of 3 μA (Figure 3a), and their diameters were 50–100 nm. Ordered AgNWs were obtained when prepared at an *I* of 5 μA (Figure 3b), and their diameters were also 50–100 nm. A dense silver nanoarray parallel to the substrate surface was formed when the *I* was 8 μA (Figure 3c), and the diameters of individual strands were 80–120 nm. When the *I* was 12 μA, the resulting nanostructure was a compact AgNB cluster arranged perpendicular to the substrate surface, and its surface roughness was larger than those of the earlier morphologies (Figure 3d).

Figures 3a–d show that silver nanostructures with different morphologies were obtained by changing the *I* used during synthesis. This was because the crystal nucleation rate and accumulation mode of the silver atoms changed with different *I*. Silver ions diffused to the cathode and formed at the interface during the initial stage of applied DCEF. When the *I* was 3 μA, the DCEF was very weak, so the AgNW growth direction was random. When the *I* was increased to >5 μA, the DCEF was stronger, and silver ions tended to migrate along the DCEF direction. Therefore, nanowires grew parallel to the DCEF. Increasing the DCEF strength resulted in many silver ions accumulating at the interface. This caused the nucleation of silver atoms in a supersaturated state. Thus, the morphology of the silver nanostructures changed significantly with the *I* used during synthesis. Disordered AgNWs, ordered AgNWs, Ag nanoarrays and AgNB clusters could be obtained by changing the *I*. Figure 3e shows a typical energy dispersive spectrum (EDS) of these nanostructures, and all peaks indicated that the nanostructures consisted of pure silver.

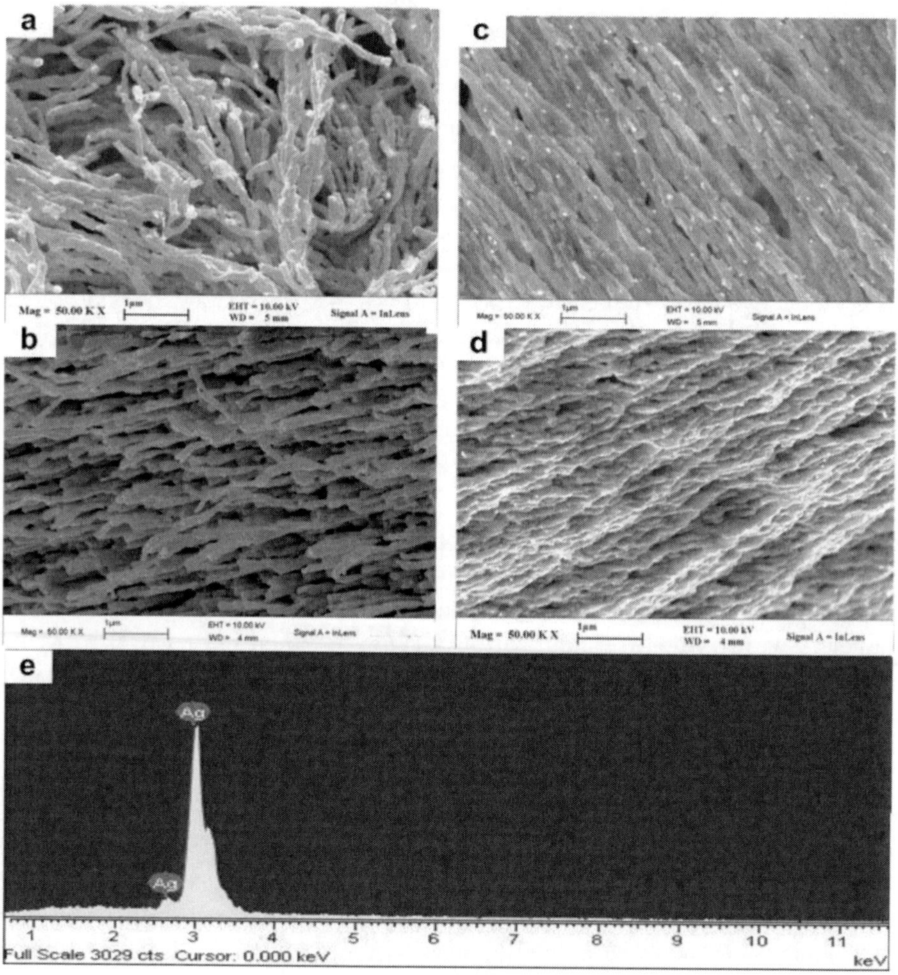

Figure 3. a)–d) SEM images of silver nanostructures prepared at *I* of (a) 3, (b) 5, (c) 8 and (d) 12 μA. The scale bars are 1 μm [25]. e) EDS spectrum of all nanostructures with different morphologies (i.e. for all samples show the same spectrum).

2.3. Preparation of AgNP Substrates by Vacuum Thermal Evaporation

AgNP films were deposited on smooth clean silicon wafers by vacuum thermal evaporation. Several 8×10 mm^2 silicon wafers were first cut from a large piece of silicon. They were then ultrasonically cleaned, and placed in a vacuum coating sample holder. The evaporation of silver began when the vacuum chamber pressure reached ~1×10^{-3} Pa. The sample holder was rotated at 20 rpm during deposition. The deposition rate was ~10 nm per second, and the total deposition time was ~10 seconds. Figure 4 shows an atomic force microscopy (AFM) image of the as-prepared AgNP film. The nanoparticle film largely consisted of particles of ~40 nm in size.

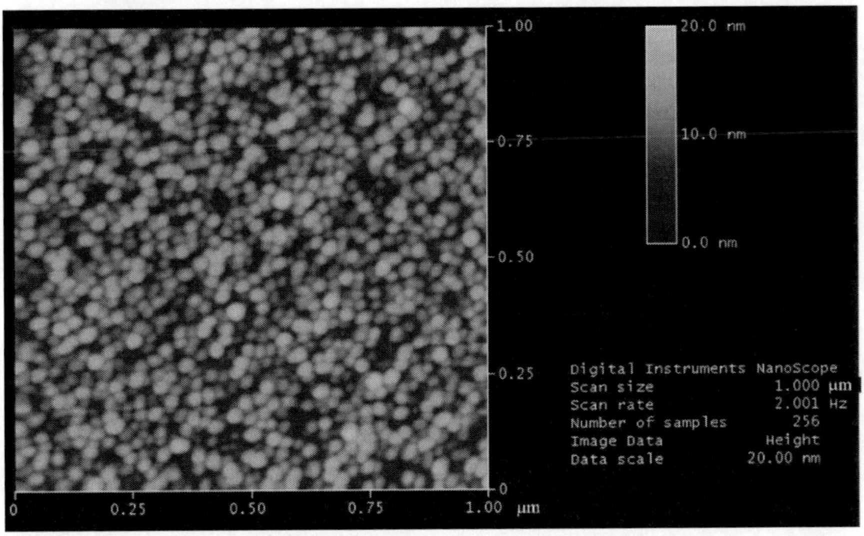

Figure 4. AFM image of the AgNP film.

3. Surface Plasmon Polaritons (SPPs) and Surface Enhancement Effects

3.1. SPP Excitation in the Ag Nanostructures

We investigated the excitation of SPPs on the AgNW surface, and studied the interaction between the SPPs and electrons. Photoinduced SPPs had an appreciable influence on electron transport in the AgNWs. The excitation efficiency of SPPs on the AgNW surface was also larger than those on a flat Ag surface [26].

An AgNW bundle was fabricated by the improved solid-state ionics method described earlier. Figure 5 shows a photograph of the AgNW bundle. The sample was placed in a cryostat at ~1×10^{-3} Pa, to control its temperature and prevent it being affected by air (Figure 6). An optical chopper was used to periodically interrupt a 532-nm monochromatic beam from a semiconductor laser. The transmission coefficient at 532 nm of the cryostats quartz glass window was ~90%. To avoid generating a photoinduced voltage, the laser spot was located near the geometric center of the AgNW bundle [24] (Figure 5).

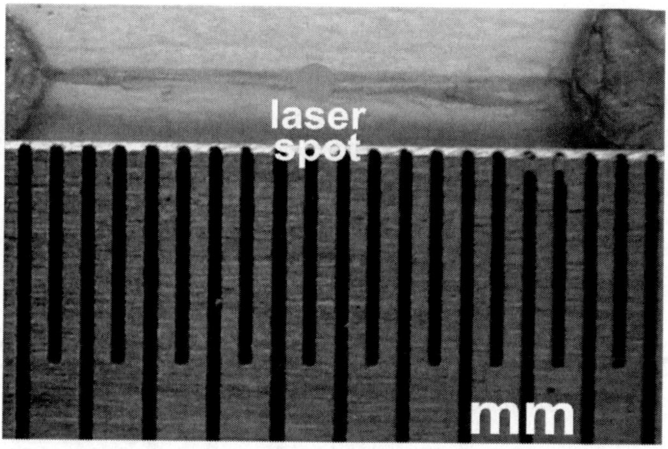

Figure 5. Photograph of the as-fabricated AgNW bundle located between two conducting silver paint electrodes.

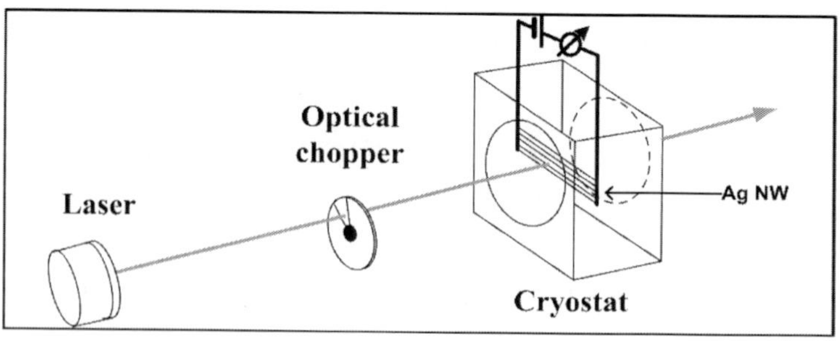

Figure 6. Schematic of the setup for measuring the photoconductivity of the AgNW bundle [26].

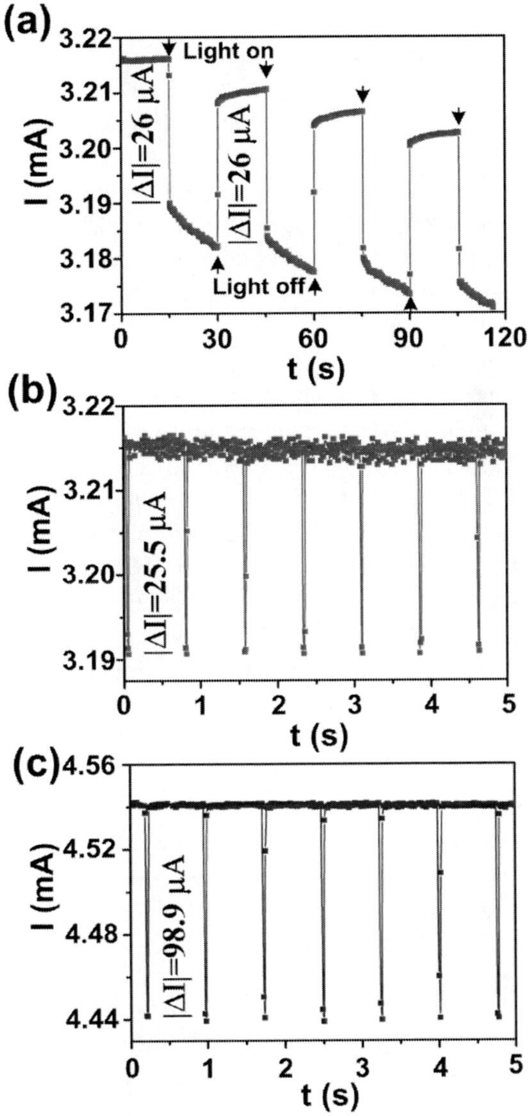

Figure 7. (a) Dynamic response of negative photoconductivity at room temperature when the AgNW bundle was irradiated (on/off circles) by a 142.3 mW continuous laser, under a bias voltage of 0.1 V. (b) and (c) show dynamic responses of negative photoconductivity at room temperature and liquid-nitrogen temperature, respectively, when the AgNW bundle was irradiated by 142.3 mW pulsed light (pulse duty factor ~6%), under a bias voltage of 0.1 V [26].

The dynamic response of negative photoconductivity was investigated at room temperature, as shown in Figure 7a. The AgNW bundle was irradiated (on/off circles) by a 142.3 mW continuous laser, under a bias voltage of 0.1 V. A significant decrease in the current of the AgNW bundle was observed. The response speed was much faster than that of the subsequent further current decrease. When the light was turned on, the current immediately decreased from 3.216 to 3.190 mA, and then slowly decreased from 3.190 to 3.182 mA. When the light was turned off, the current immediately increased from 3.182 to 3.208 mA, and then slowly increased from 3.208 to 3.211 mA. For every on/off circle, the fast response process resulted from the photoinduced SPP-electron interactions, and the slow response process resulted from SPP-induced heating. Both SPP-electron scattering and SPP-induced heating contributed to photoconductivity.

SPP-electron scattering was the dominant effect for short time scales, i.e. for light with a small pulse duty factor. To avoid any influence on the negative photoconductivity of the AgNW bundle by Joule and/or SPP-induced heating, the conductive properties of the AgNW bundle were investigated under a small bias voltage of 0.1 V, and a pulsed laser field with a small pulse duty factor. In short, the dynamic responses of negative photoconductivity for the AgNW bundle illuminated by pulsed light were studied. Figure 7b and c show the dynamic responses at room temperature and liquid-nitrogen temperature, respectively.

Figure 7c shows that the dark current (I_d) was constant at 4.542 mA at liquid-nitrogen temperature. Figure 7b shows that at room temperature, I_d fluctuated around 3.216 mA, over the 5-second measurement time scale. When the light was turned on, a significant decrease in the current of the AgNW bundle was observed. Comparing Figure 7a, b and c indicated that the fast SPP-electron scattering process made the biggest contribution to the current change, at short time scales. This transmission of the AgNW bundle was attributed to the SPP-electron interaction on the AgNW surface.

The effective excitation of SPPs depends on the surface roughness [27]. In support of this, we deposited Ag thin film strips of thickness 50–60 nm and width 1 mm on the optically flat surface of an insulating substrate, by vacuum thermal evaporation (Figure 8a). The two ends of strips were connected by conductive silver paint electrodes. When the current for the Ag film was measured under similar conditions, no obvious change was observed upon irradiation (Figure 8c). Depositing ultrasonically dispersed AgNWs on the Ag film under the laser spot (Figure 8b) increased the surface roughness, and resulted in a decreased current upon irradiation (Figure 8d). This illustrated the importance of surface roughness, and indicated that the excitation efficiency of SPPs on the AgNW surface was larger than on a flat Ag film.

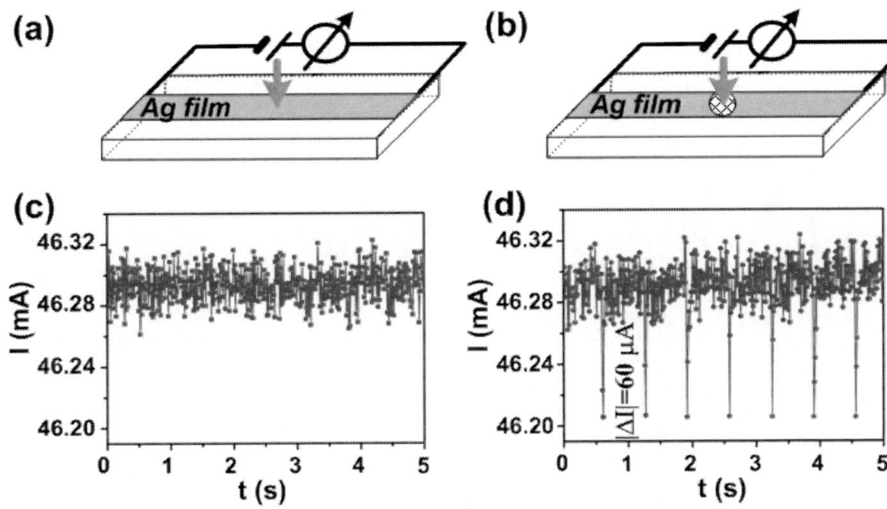

Figure 8. (a)–(b) Schematics of Ag thin film strips without and with deposited AgNWs, respectively. (c)–(d) Dynamic responses of negative photoconductivity at room temperature for the samples in (a)–(b), respectively, when the Ag films were irradiated by 142.3 mW pulsed light (pulse duty factor ~6%), under a bias voltage of 0.1 V [26].

3.2. SERS Effect of the Silver Nanostructure Substrates

As described in section 2.2, the morphology of the silver nanostructures differed when prepared at differing *I*. A smooth metal surface cannot induce significant SERS. Only an appropriate metal surface with a coarse interface can be used in high-sensitivity SERS spectroscopy. R6G solutions of concentration 1×10^{-13} mol L^{-1} were used to investigate the Raman enhancement effect of the silver nanostructured substrates. Silver nanostructures prepared at different *I* were used as SERS substrates, and their Raman spectra are shown in Figure 9.

At an *I* of 3 μA, the prepared disordered AgNWs only detected the intrinsic Raman envelope peak of R6G (Figure 9a), because of their disordered arrangement (Figure 3a). The ordered AgNWs and AgNB cluster prepared at *I* of 5 (Figure 3b) and 12 (Figure 3d) μA exhibited very strong SERS effects, as shown in Figure 9b and d, respectively. The intrinsic Raman peaks of R6G were observed at 612, 773, 1128, 1184, 1310, 1361, 1506, 1573, 1597 and 1650 cm^{-1}. The Ag nanoarray prepared at an *I* of 8 μA (Figure 3c) detected all the intrinsic Raman peaks of R6G, but individual Raman peaks at 1506, 1573, 1597 and 1650 cm^{-1} were not resolved (Figure 9c). Thus, the use of ordered AgNWs and the AgNB cluster as SERS substrates detected all Raman signals of 1×10^{-13} mol L^{-1} R6G. The orderly arrangement of ordered AgNWs and the roughness of the AgNB cluster resulted in both nanostructures exhibiting significant SERS effects.

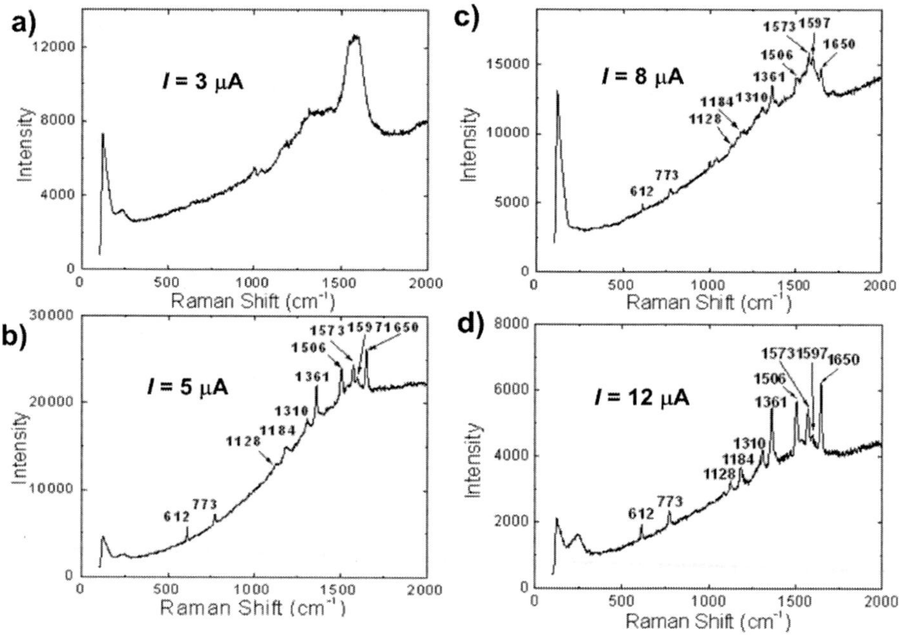

Figure 9. Raman spectra of 1×10^{-13} mol L^{-1} R6G solutions on various silver nanostructured substrates, which were grown under different *I* [25].

The Raman detection limit of R6G when using silver colloid particles as a SERS substrate is reportedly 2×10^{-11} mol L^{-1} [19]. The Raman detection limit of R6G when using monolayer AgNWs as a SERS substrate is reportedly 1×10^{-9} mol L^{-1} [28]. In the current study, the ordered AgNWs and AgNB cluster could detect all the Raman signals of 1×10^{-13} mol L^{-1} R6G.

To explore the R6G detection limit, 1×10^{-14}, 1×10^{-15} and 1×10^{-16} mol L^{-1} R6G solutions were used to assess the Raman enhancement effect of the ordered AgNWs and AgNB cluster substrates. Figure 10a shows that when ordered AgNWs were used as a SERS substrate, only envelopes of the intrinsic R6G Raman peaks at 1560 and 1590 cm^{-1} (blue) were apparent for 1×10^{-14} mol L^{-1}

R6G. Therefore, the R6G detection limit as measured by the ordered AgNW SERS substrate was 1×10^{-13} mol L^{-1} (red). Figure 10b shows that for the 1×10^{-14}, 1×10^{-15} and 1×10^{-16} mol L^{-1} R6G solutions, all intrinsic R6G Raman peaks were observed when the AgNB cluster was used as the SERS substrate. Thus, the AgNB cluster SERS substrate could detect all the Raman signals of 1×10^{-16} mol L^{-1} R6G.

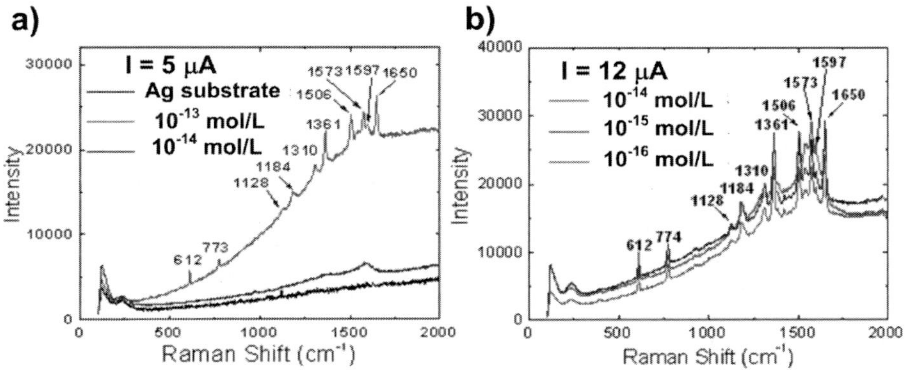

Figure 10. Raman spectra of R6G solutions with different concentrations detected by (a) ordered AgNWs and (b) the AgNB cluster as SERS substrates [25].

4. Influence of Silver Nanostructure Surface Oxidation on the Surface Enhancement Effect

The oxidation of silver nanostructure surfaces readily occurs upon their exposure to air. This is especially so for the rough surfaces of nanostructures, whose surface area to volume ratios are very large. We now explore the influence of AgNP film oxidation on the surface enhancement effect.

4.1. Change in AgNP Film Thickness and Surface Morphology after Different Oxidation Times

AgNP films were prepared by vacuum thermal evaporation, as described in section 2.3. To measure the AgNP film thickness, we made a mask at the center of the silicon wafer. Deposition during vacuum thermal evaporation did not occur under the mask. The mask was removed after deposition, and AFM was used to scan the film surface and measure the AgNP film thickness. Multiple points near the film border were measured and averaged, to obtain an AgNP film thickness of 131.4 nm. The same method was used to measure the thickness of the AgNP film after its oxidization in air. The change in the AgNP film thickness (*Δh*) after different oxidation times was defined as the thickness after oxidation minus the original thickness of 131.4 nm. Figure 11 shows *Δh* with increasing oxidation duration.

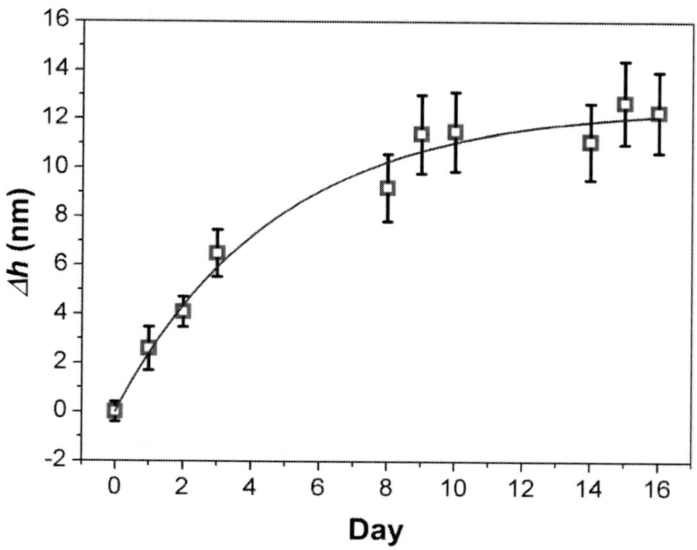

Figure 11. Change in AgNP film *Δh* with increasing oxidation duration.

Figure 11 shows that at 0–3 days, the increase in *Δh* with increasing duration was significant. The increase in *Δh* became gentler after oxidization for >10 days. *Δh* finally stabilized to ~12 nm. At the beginning of oxidation, AgNPs on the surface of the film were thought to be closely distributed, which favored their reaction with oxygen. The film surface then oxidized to Ag_xO. More and more oxygen reacted with the AgNPs with increasing time, leading to an increase in the volume of Ag_xO. This increase in the volume of Ag_xO resulted in the increase in AgNP *Δh*. At >10 days, a compact Ag_xO layer had formed and covered the AgNP film surface, preventing any further oxidation of Ag particles within the film. Therefore, the increase in *Δh* slowed and stabilized to a final value of ~12 nm. During the oxidation of Ag atoms, Ag_xO reportedly reversibly decomposes into

Ag and oxygen atoms [29]. Thus, Δh should be expected to change with conditions such as temperature, humidity and pressure.

The Δh oxidation data were fitted to $\Delta h = A(1 - e^{-t/\tau})$, where A = 12.55 ± 0.58 nm and τ = 4.71 ± 0.53 days. The fitted curve is shown (blue) in Figure 11.

We investigated the change in AgNP film surface morphology, particle size and roughness with oxidation time. When the AgNP film was first deposited, we removed it from the vacuum chamber and immediately scanned its surface by AFM. Twenty surface positions were randomly selected for scanning, each with an area of 1 μm^2. This process was repeated every day during the 2 weeks of oxidation, yielding 300 images.

Figure 12 shows the change in AgNP film surface morphology, during oxidation for 0–14 days. A representative image recorded from each day is shown. At day 0, Ag particles were spheres with uniform particle sizes. After day 2, Ag particles were spheres, but their size had increased. After day 4, Ag particles began exhibiting irregular shapes. Some protrusions appeared on particle surfaces, and particles gradually divided into smaller particles with increasing oxidation time. This resulted in a significant increase in the number of Ag particles on the film surface.

The average number of particles per μm^2 (n) and Ag particle size distribution were obtained by statistical analysis of the 300 images, and are shown in Figure 13a and b, respectively. Figure 13a shows that the number of Ag particles on the surface per μm^2 increased with increasing oxidation time, especially after 3 days. Oxidation resulted in Ag particles gradually dividing into a large number of smaller particles. The number of Ag particles larger than 60 nm was low, and AgNPs of size 20–40 nm were the most abundant. The 20–40-nm particles were oxidized. Some oxidized homogeneously, increasing their size and thus the proportion of AgNPs of size 40–60 nm. Some oxidized heterogeneously, dividing into smaller particles and thus increasing the proportion of AgNPs of size 20–40 nm. Overall, increasing oxidation time decreased the amount of 20–40 nm particles (red) and increased the amount of 40–60 nm particles (blue). The amount of particles smaller than 20 nm (black) initially increased and then fluctuated with increasing oxidation duration. A statistical analysis of the AgNP film roughness is shown in Figure 13c.

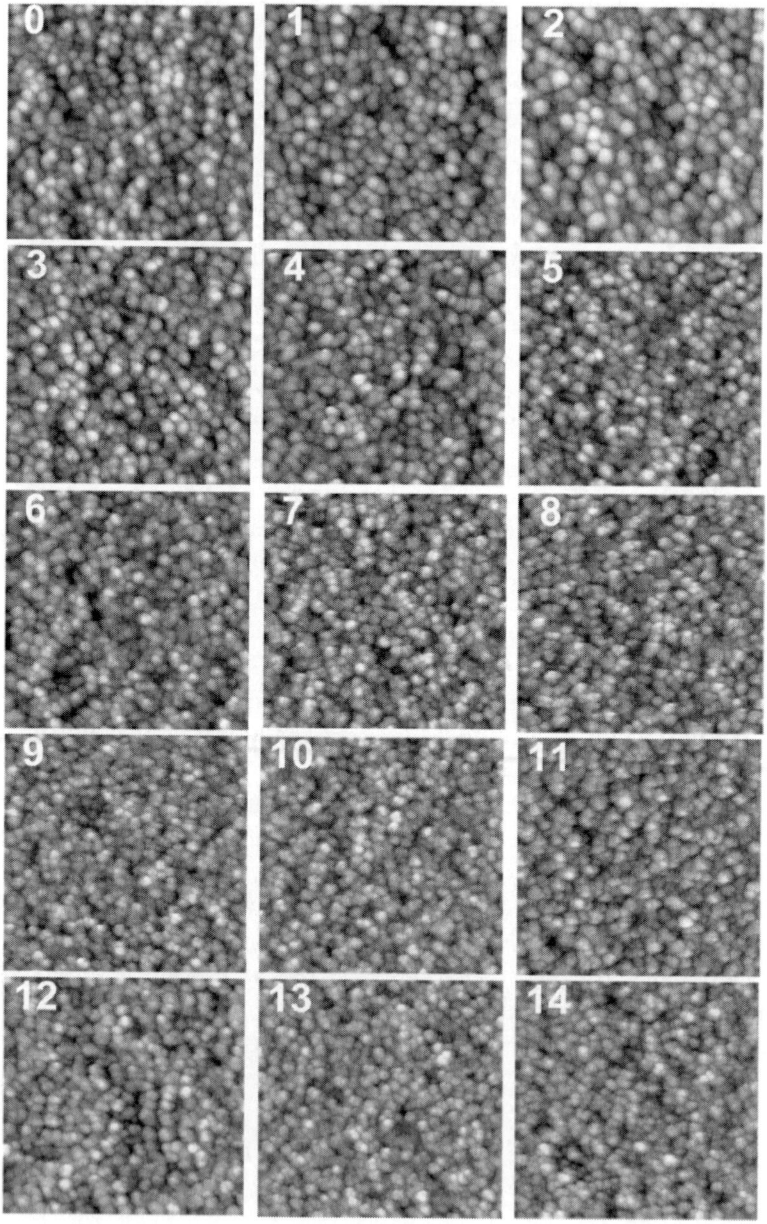

Figure 12. AFM images of the AgNP film surface morphology after oxidation for increasing numbers of days. The areas of the images were all 1 μm^2.

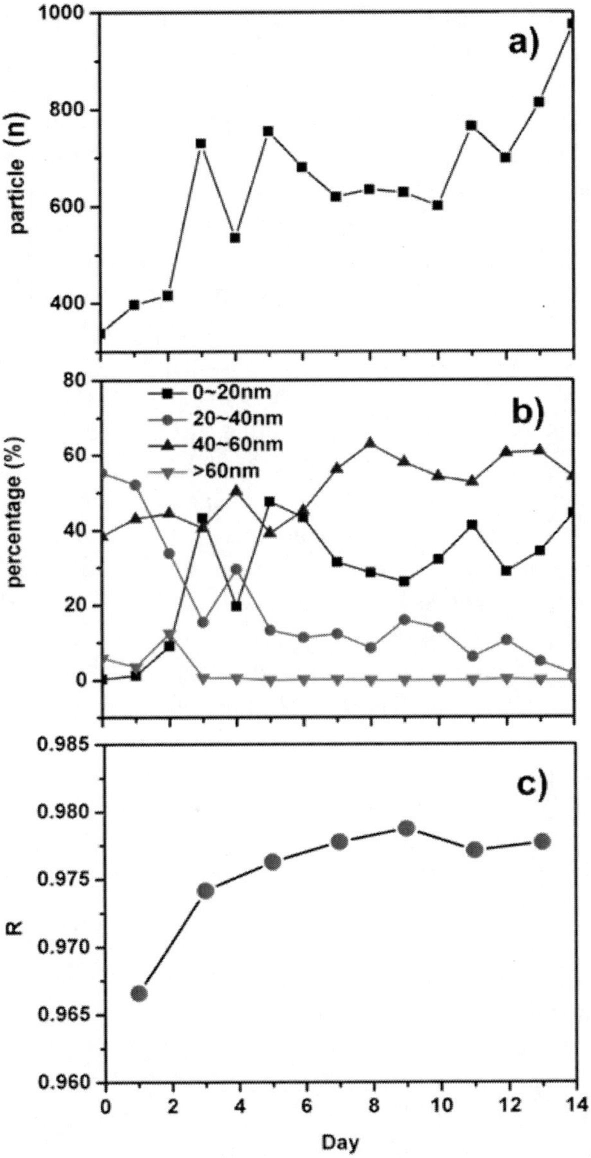

Figure 13. Changes in (a) average number of particles per μm^2, (b) Ag particle size distribution and (c) AgNP film roughness with increasing oxidation duration.

4.2. Influence of Oxidation Time on the Surface Enhancement Effect of AgNP Films

AFM was used to measure the change in AgNP film thickness of a further 15 films, prepared as described in section 2.3. These AgNP films were stored in a vacuum chamber, and one film was removed every 24 hours and allowed to oxidize in air. AgNP films oxidized for 0–14 days were therefore obtained 2 weeks later, and were used as surface enhancement substrates. A R6G concentration of 1×10^{-6} mol L^{-1} was used to investigate the enhancement effect of the AgNP film substrates. The detection of R6G on the AgNP substrates was carried out using a confocal Raman spectrometer (Renishaw RM2000, UK). The excitation source was a focused Ar^+ laser (wavelength 514 nm, focal area ~2×2 μm with a 50× objective). The exposure time was 20 seconds and the detection range was 500–4000 cm^{-1}. The laser power was 0.47 mW. We prepared an AgNP sample in advance, and recorded its spectra under the above conditions for 12 minutes. The spectra were constant, showing that the sample was stable. Each sample required an actual spectral collection time of 6 minutes. Thus, laser irradiation during the short detection times did not affect the AgNP film surface oxidation state.

The detection results for substrates oxidized for 0–14 days are shown in Figure 14a–c. A change in spectral features with increasing oxidation was apparent. Figure 14d shows that there was a change in the relative intensity of the Raman peak at 1648 cm^{-1} with increasing oxidation duration. The result was normalized to the intensity of this peak at day 0. Figure 14d shows that the Raman intensity undulated with increasing oxidation duration. The Raman intensity of R6G on the oxidized AgNP substrate was about 50 times that of R6G on the unoxidized AgNP substrate.

Raman scattering intensity is proportional to the square of the dipole moment (P) induced in the molecule by the incident field. $P = \alpha \cdot E$, where α is the molecular polarizability and E the incident field. SERS largely contributes to signal intensity by a change in electric field experienced by the molecule, and a change in molecular polarizability. The former is generally considered to be a physical mechanism, and results from the enhanced local electric field of surface molecules. The latter is a chemical mechanism, which considers that the reaction between the metal and molecule changes the molecular polarizability [30-32]. The influence of the physical mechanism is reportedly ~2–4 orders of magnitude stronger than the chemical mechanism [30-32], so the physical mechanism is the major contributor.

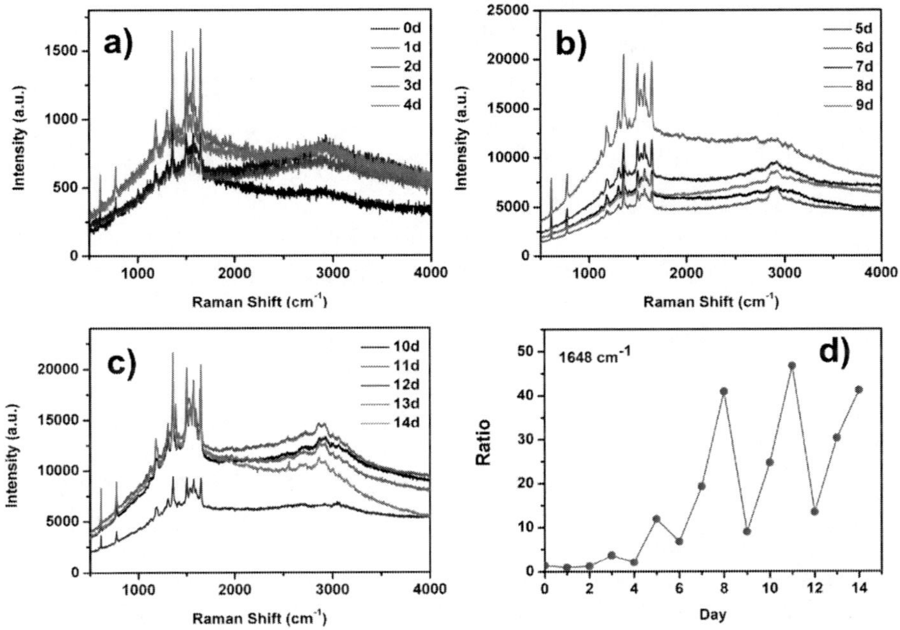

Figure 14. (a)–(c) SERS spectra for 1×10^{-6} mol L^{-1} R6G on AgNP substrates oxidized for different durations. (d) Relative intensity of the Raman peak at 1648 cm^{-1} with increasing oxidation duration.

SEF is a complex process, and has been the subject of long study. In SEF theory, molecular fluorescence can be enhanced or quenched, as a result of the interaction/ competition of several mechanisms. These include the enhancement of the local electric field near the metal surface, radiative energy transfer between the molecule and metal, non-radiative energy transfer between the molecule and metal, and oxygen vacancies [33-36]. These mechanisms are affected by the intensity of the local electrical field, distance between the molecule and metal surface, quantity of hot-spots, et cetera.

As mentioned above, both SERS and SEF are affected by the local field enhancement caused by the plasmon resonance on metal surface. The resonance of a plasmon exciting a single molecule is theoretically limited to a very narrow wavelength range. These electrons are loosely bound, and can freely couple with adjacent particles. Between adjacent particles, a plasma wave forms from the coupling of plasmons of adjacent particles. This enhances the local electrical field, and leads to a stronger surface enhancement effect. These places are called hot-spots, and are related to the AgNP film surface roughness, the number of nanoparticles and their size. The surface enhancement effect is affected by all these factors during oxidation.

5. Silver Nanostructure Substrates Applied in the Detection of λ-DNA

5.1. SERS of λ-DNA on Rough Silver Nanostructures

Raman spectroscopy is well suited to biomedical studies, because of its nondestructive detection, detailed molecular information, ease of sample preparation, and absence of interference by water. The surface enhancement abilities of the rough silver nanostructures were tested using λ-DNA.

The rough silver nanostructures were prepared as described in section 2.1. The *I* during the growth of the AgNBs was 12 μA, which yielded a dense unidirectional morphology (Figure 15a).

An Alpha-300SR scanning confocal Raman spectrometer (WITec, Germany) was used to examine λ-DNA on the AgNB substrates. The excitation source was a focused Ar^+ laser (excitation wavelength 532 nm, focal area ~360×360 nm with a 100× objective). The laser power was 0.5 μW, the exposure time was 100 ms and the detection range was 400–2000 cm^{-1}. AgNBs were fixed on a clean slide as the SERS substrate, and the λ-DNA concentration was 50 $ng \cdot \mu L^{-1}$. The focal area was very small, so a relatively high λ-DNA concentration was used. The λ-DNA solution was dried before detection.

A 25×16 μm region of a dry λ-DNA sample was scanned, and the resulting Raman spectral image of 100×64 pixels is shown in Figure 15b. Each pixel had a corresponding Raman spectrum, and the gray level of each pixel was determined by the intensity of its corresponding spectrum. There was either no λ-DNA or no λ-DNA was close enough to the AgNB substrate to cause SERS in the detection area, for the dark pixels such as A in Figure 15b. Bright pixels including B, C, D and E exhibited strong SERS signals. The adsorption of λ-DNA on the AgNB substrate was affected by the rough surface (Figure 15a). Thus, the intensities of each pixel differed, rendering the Raman image into many dots of varying brightness.

The spectra of pixels A, B, C, D and E are shown in Figure 16. Peaks common to dry λ-DNA occurred at 1245 and 1380 cm^{-1}. The peak at 1245 cm^{-1} peak was due to thymine, cytosine and adenine (T, C, A). The peak at 1380 cm^{-1} was due to thymine, adenine and guanine (T, A, G). Tentative assignments for peaks in the spectra B, C, D and E are given in Table 1 [37-42]. The Raman peaks reflected the backbone deformations of dry λ-DNA. The laser used had a small focal size of 360 nm. It is possible that only part of a single λ-DNA molecule resided in the focal area, and that its Raman peaks revealed structural information for local regions within that single molecule.

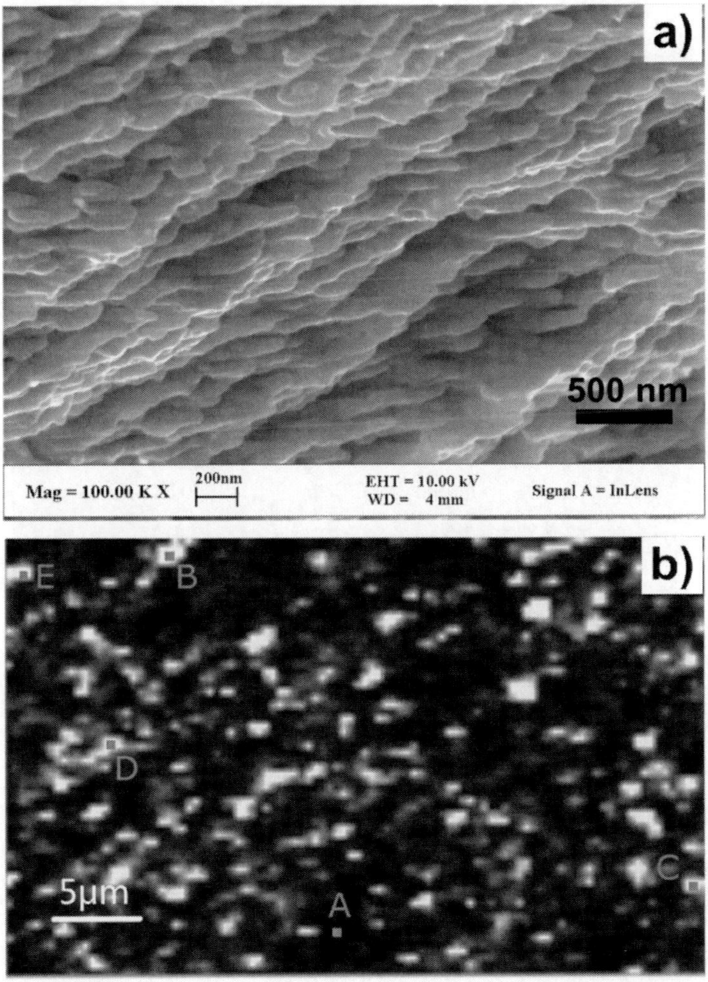

Figure 15. (a) SEM image of the AgNB structure. AgNB was rough, dense and unidirectional, and composed of compacted 100 nm sized units like nanoscale buds. (b) Scanning confocal Raman spectral image of dry λ-DNA on the AgNB substrate. The gray level of each pixel was determined by the intensity of its corresponding spectrum.

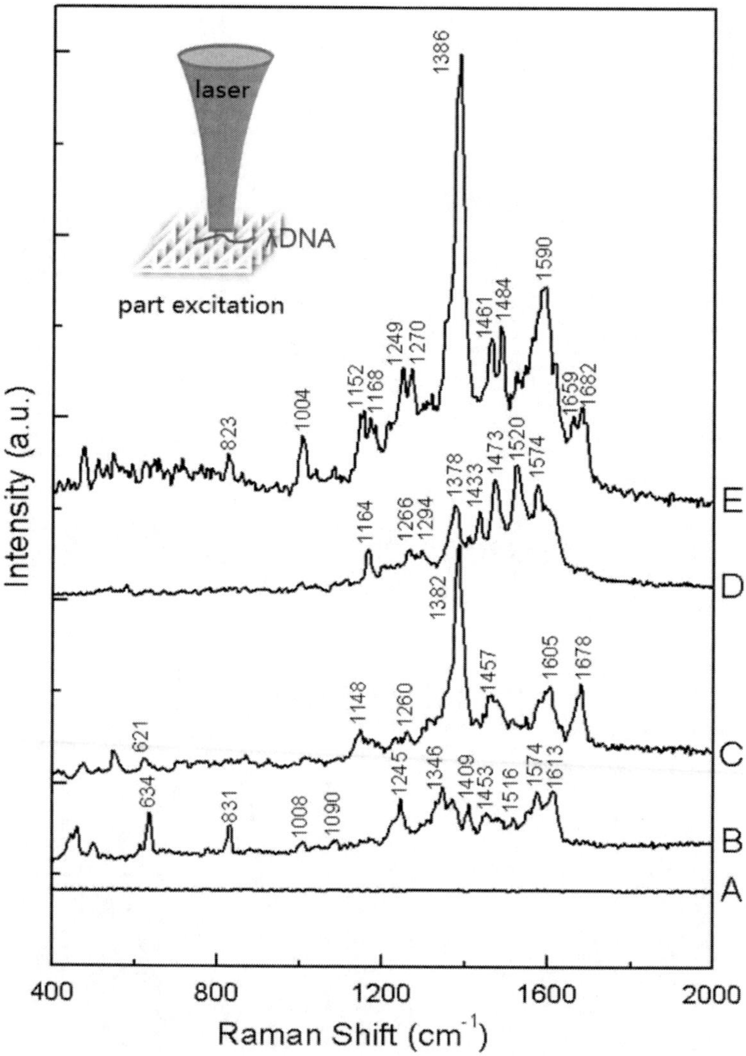

Figure 16. SERS spectra of the pixels A, B, C, D and E in Figure 15b. Some of the peaks and corresponding structural information in B–E reflected the local conformation of λ-DNA. The inset shows a schematic of the excitation of a λ-DNA molecule only partly residing in the focal area.

Table 1. Tentative assignments for SERS spectra of λ-DNAs on AgNB substrate

λ-DNA concentration(50 ng / μL)					Tentative Assignmen
A	B	C	D	E	
	634	621			G
	831			824	PO_2^-
					Deoxyribose
	1008			1004	Deoxyribose
	1090				PO_2^-
		1148	1164	1152, 1168	Deoxyribose
	1245	1260	1266	1249, 1270	T,C,A
			1294		C,A
	1346				A,G
	1409	1382	1378	1386	T,A,G
	1453	1457	1433, 1447	1461, 1484	Deoxyribose, CH_2
	1516		1520		A
	1574, 1613	1605	1574	1590	A,G
		1678		1659, 1682	T

[a]Abbreviations:A, adenine; G, guanine; C, cytosine; T, thymine.

The spectra of B–E in Figure 16 exhibited very high signal-to-noise ratios (SNRs), and thus contained highly resolved structural information. SPR theory states that the Raman and fluorescence signals should both have been enhanced, when λ-DNA was close to the AgNB substrate. The enhancement factor increased as the distance between λ-DNA and the substrate decreased. However, fluorescence would have been quenched if λ-DNA was completely adsorbed on the AgNB substrate [43-45]. Dry λ-DNA was strongly adsorbed on the AgNB substrate, leading to a low intensity of the fluorescence background in spectra B-E, but with much stronger Raman peaks. This resulted in higher SNRs, and revealed subtle structural information. Thus, the AgNB substrate was used in nanoscale localized scanning confocal Raman spectroscopy, to obtain high SNR Raman spectra with abundant structural information. This revealed localized conformations of single λ-DNA molecules, and is a promising approach for the diagnosis of single chromosomes.

5.2. SEF of λ-DNA on AgNB

The substrate used was the same as that shown in Figure 15a. AgNBs were fixed on a clean slide, and used as a SEF substrate. λ-DNA was labeled with YOYO-1, and the λ-DNA concentration was 2×10^{-14} mol L^{-1}. 10 μL of λ-DNA and 10 μL of tris-HCl buffer solution were added to the AgNBs. The sample was excited with an Ar^+ laser (wavelength 488 nm), and detected with a fluorescence microscope.

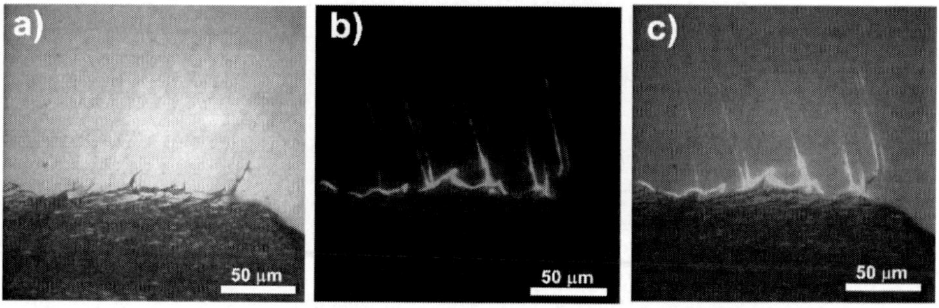

Figure 17. (a) Transmission optical image of unoxidized AgNB. (b) Fluorescence image of λ-DNA on unoxidized AgNB. (c) Overlap of images in Figures 17a and b.

The results for the unoxidized and oxidized substrates are shown in Figures 17 and 18, respectively. Figure 17a shows the image of the sample exposed in transmission light, and Figure 17b shows the fluorescence image of the sample excited at the same place. Figure 17c shows the overlap of the images in Figure 17a and b. The fluorescence was quenched on the AgNB. However, the fluorescence intensity along the edge of the AgNB was enhanced greatly compared with that of λ-DNA. Fluorescence enhancement was observed at more than 10 μm from the edge of the AgNB. This indicated that fluorescence enhancement occurred over a relatively long distance. Figure 18a shows the image of the sample exposed in transmission light, and Figure 18b shows the fluorescence image of the sample excited at the same place. In contrast to the result in Figure 17, the fluorescence of λ-DNA on AgNB was not quenched. This was because non-radiative energy transfer is proportional to the third power of the distance between the molecule and surface. Thus, non-radiative energy transfer led to the fluorescence quenching [35], because λ-DNA was in direct contact with the unoxidized metal substrate. However, λ-DNA was separated by a 10-nm-thick oxidation layer on oxidized AgNB, as discussed in section 4.2. Thus, fluorescence was quenched by the unoxidized AgNB, but not by the oxidized AgNB.

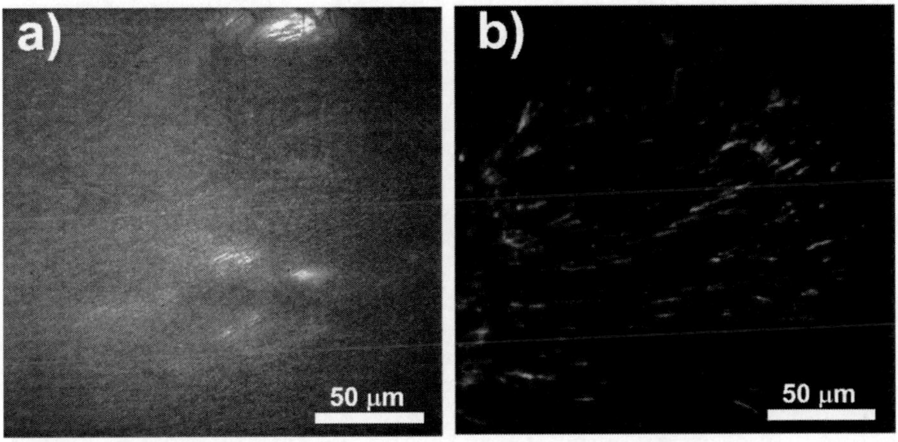

Figure 18. (a) Transmission optical image of oxidized AgNB. (b) Fluorescence image of λ-DNA on oxidized AgNB.

Conclusion

This chapter introduced an improved solid-state ionics method combined with vacuum thermal evaporation to synthesize AgNW, AgNB and AgNP structures. SPPs excited by an optical field on a rough Ag film surface were stronger than those on a smooth Ag film surface. SPPs on the AgNW surface were stronger than those on the Ag film.

We experimentally and theoretically explored the influence of AgNB oxidation on the surface enhancement effect. This influence is important, given that silver is easily oxidized in air. The Raman intensity of R6G on an oxidized AgNP substrate was ~50 times greater than that on an unoxidized substrate.

The surface enhancement abilities of several silver nanostructures were discussed and compared. AgNB was an excellent substrate for surface enhancement detection, so was applied to detect λ-DNA. Using an AgNB substrate, the Raman signal of λ-DNA obtained single molecular detection limits, which revealed information on the localized conformation of a single DNA molecule. The fluorescent signal of λ-DNA was enhanced on the AgNB substrate, and was detected at more than 10 μm from the AgNB edges. This indicated that the fluorescence enhancement occurred over a relatively long distance. AgNB therefore has potential in the diagnosis of single chromosomes, and the high-sensitivity detection of biomacromolecules.

Acknowledgments

This work was partially supported by the National Science Foundation of China (Nos. 11374172 and11374180) and the Specialized Research Fund for the Doctoral Program of Higher Education (No. 20120002110024).

References

[1] Fleischmann M, Hendra P J, McQuillan, Raman spectra of pyridine adsorbed at a siler electrode, *Chem. Phys. Lett.* 1974, 26(2):163-166

[2] Jeanmaire D L, Van Duyne R P., Surface Raman spectroelectrochemistry: Part I. Heterocyclic, aromatic, and aliphatic amines adsorbed on the anodized silver electrode, *J. Electro. Anal. Chem.* , 1977, 84(1):1-20

[3] Albrecht M G, Creighton J A, Anomalously intense Raman spectra of pyridine at a silver electrode. *J. Am. Chem. Soc.*, 1977, 99(15):5215-5217

[4] You Y M, Purnawirman N A, Hu H L, et al, Tip-enhanced Raman spectroscopy using single-crystalline Ag nanowire as tip, *J. Raman Spectroscopy*, 2010, 41(10):1156-1162

[5] Wang X T, Shi W S, She G W, et al., High-performance surface-enhanced Raman scattering sensors based on Ag nanoparticles-coated Si nanowire arrays for quantitative detection of pesticides, *Appl. Phys. Lett.,* 2010, 96: 053104

[6] Camargo P H C, Cobley C M, Rycenga M, et al., Measuring the surface-enhanced Raman scattering enhancement factors of hot spots formed between an individual Ag nanowire and a single Ag nanocube, *Nanotechnology,* 2009, 20: 434020

[7] Fang Y R, Wei H, Hao F, et al., Remote-Excitation Surface-Enhanced Raman Scattering Using Propagating Ag Nanowire Plasmons, *Nano Lett.,* 2009, 9(5):2049-2053

[8] Deng S, Fan H M, Zhang X, et al., An effective surface-enhanced Raman scattering template based on a Ag nanocluster–ZnO nanowire array, *Nanotechnology,* 2009, 20: 175705

[9] Emmanuel F, Samuel G, Surface enhanced fluorescence, *J. Phys. D, Appl. Phys.*, 2008, 41: 013001

[10] Lu F T, Zheng H R, Fang Y, Studies of surface-enhanced fluorescence, *Progr. in Chem*, 2007, 19: 256-266

[11] Lakowicz J R, Radiative decay engineering: biophysical and biomedical applications, *Anal. Biochem.*, 2001, 298(1): 1-24

[12] Lakowicz J R, Radiative decay engineering: metal-enhanced fluorescence and plasmon emission, *Anal. Biochem.*, 2005, 337(2): 171-194

[13] Huang X H, El-Sayed I H, et al., Cancer Cells Assemble and Align Gold Nanorods Conjugated to Antibodies to Produce Highly Enhanced, Sharp, and Polarized Surface Raman Spectra: A Potential Cancer Diagnostic Marker, *Nano Lett.* 2007, 7(6): 1591-1597

[14] Sun Y H, Liu K, Miao J, et al., Highly Sensitive Surface-Enhanced Raman Scattering Substrate Made from Superaligned Carbon Nanotubes, *Nano Lett.,* 2010, 10(5): 1747-1753

[15] Lee S J, et al., Polarization-Dependent Surface-Enhanced Raman Scattering from a Silver-Nanoparticle-Decorated Single Silver Nanowire, *Nano Lett.,* 2008, 8(10): 3244-3247

[16] Wei H, Hao F, et al., Polarization Dependence of Surface-Enhanced Raman Scattering in Gold Nanoparticle–Nanowire Systems, *Nano Lett.,* 2008, 8(8): 2497-2502

[17] Ojeda J F, Xie C G, et al., Chromosomal analysis and identification based on optical tweezers and Raman spectroscopy, *Opt. Express,,* 2006, 14: 5389-5393

[18] Kneipp K, Wang Y, Kneipp H, et al. Single molecule detection using surface-enhanced Raman scattering (SERS), *Phys Rev Lett,* 1997, 78: 1667-1670

[19] Nie S M, Emery S R., Probing single molecules and single nanoparticles by surface-enhanced Raman scattering, *Science,* 1997, 275: 1102-1106

[20] Lee C H, Tian L M, Abbas A, et al., Directed assembly of gold nanorods using aligned electrospun polymer nanofibers for highly efficient SERS substrates, *Nanotechnology,* 2011, 22: 275311

[21] Castillo F, De la Rosa E, Perez E, Gold aggregates on silica templates and decorated silica arrays for SERS applications, *Eur. Phys. J. D*, 2011, 63: 301-306

[22] Parfenov A, Gryczynski I, Malicka J, et al., Enhanced fluorescence from fluorophores on fractal silver surfaces, *J. Phys. Chem.,* 2003, 107: 8829-8833

[23] Sun J L, Zhang J H, Liu W, Liu S, Sun H S, Jiang K L, Li Q Q and Guo J H, Shape-controlled synthesis of silver nanostructures, *Nanotechnology*, 2005, 16: 2412-2414

[24] Xu J, Sun J L and Zhu J L, Thermo- and photoinduced voltages in Ag heterodimensional junctions, *Appl. Phys. Lett.,* 2007, 91: 161107

[25] Xu D P, Dong Z M and Sun J L, Fabrication of high performance surface enhanced Raman scattering substrates by a solid-state ionics method, *Nanotechnology,* 2012, 23: 125705

[26] Sun J L, Zhang W, Zhu J L and Bao Y, Negative photoconductivity induced by surface plasmon polaritons in Ag nanowire macrobundles, *Opt. Express,* 2010, 18(5): 4066-4073

[27] Zayats A V, Smolyaninov I I and Maradudin A A, Nano-optcs of surface plasmon polaritons, *Phys. Rep.*, 2005, 408(3-4): 131-314

[28] Tao A, Kim F, Hess C, Goldberger J, He R R, Sun Y G, Xia Y N and Yang P D, Langmuir–Blodgett Silver Nanowire Monolayers for Molecular Sensing Using Surface-Enhanced Raman Spectroscopy, *Nano Lett.*, 2003, 3: 1229-1233

[29] Pettinger B, Bao X, Wilcock I, et al., Thermal-Decompisition of Silver-Oxide Monitored by Raman Spectroscopy from AgO units to Oxygen atoms Chemisorbed on the Silver Surface, *Angew. Chem. Int. Ed.*, 1994, 33: 85-86

[30] Moskovits M S., surface enhanced spectroscopy, *Rev. Mod. Phys.*, 1985, 57: 783-825

[31] Yin Y D, Surface enhanced Raman scattering theory research of molecular adsorbed on anisotropic dielectric core - shell structure metal, *J. Suzhou University,* 2011, 27: 48-52

[32] Fang J L, Wu Y, Surface enhancement Raman spectroscopy, *Surface technology*, 1994, 23: 167-194

[33] Campion A, Gallo A R, Harris C B, et al, Electronic-energy transfer to metal-surface – A test of classical image dipole theory at short distances, *Chem. Phys. Lett.*, 1980, 73: 447-450

[34] Geddes C D, Lakowicz J R, Editorial:Metal-enhanced fluorescence, *J. Fluoresc.*, 2002, 12: 121-129

[35] Kummerlen J, Leitner A, Brunner H, et al., Enhances dye fluorescence over silver island films – analysis of the distance dependence, *Mol. Phys.,* 1993, 80: 1031-1046

[36] Aslan K, Leonenko Z, Lakowicz J R, et al., Annealed silver-island films for applications in metal-enhanced fluorescence: Interpretation in terms of radiating plasmons, *J. Fluoresc.*, 2005, 15: 643-654

[37] Peticolas, W L, Raman spectroscopy of DNA and proteins, *Methods Enzymol,* 1995, 246: 389–416

[38] Deng H, Bloomfield V A, Benevides J M, Thomas G J, Dependence of the Raman signature of genomic B-DNA on nucleotide base sequence, *Biopolymers,* 1999, 50(6): 656–666

[39] Wei H, and Xu H, Surface-enhanced Raman scattering of λ-DNA, *Appl. Phys. A*, 2007, 89(2): 273–275

[40] De Gelder J, De Gussem K, Vandenabeele P, Moens L, Reference database of Raman spectra of biological molecules, *J. Raman Spectrosc.*, 2007, 38(9): 1133–1147

[41] Sun L L, Sun Y J, Xu F G, Zhang Y, Yang T, Guo C L, Liu Z L, Li Z, Atomic force microscopy and surface-enhanced Raman scattering detection of DNA based on DNA–nanoparticle complexes, *Nanotechnology*, 2009, 20: 125502

[42] Muntean, C. M.; Leopold, N.; Halmagyi, A.; Valimareanu, S., Surface-enhanced Raman spectroscopy of DNA from leaves of in vitro grown apple plants, *J. Raman Spectrosc.*, 2011, 42: 844–850

[43] Gersten J, and Nitzan A, Electromagnetic theory of enhanced Raman scattering by molecules adsorbed on rough surfaces, *J. Chem. Phys.*, 1980, 73(7): 3023–3037

[44] Kim K, Kim D J, Cho E J, Suh J S, Huh Y M, and Kim D, Nanograting-based plasmon enhancement for total internal reflection fluorescence microscopy of live cells, *Nanotechnology*, 2009, 20: 015202

[45] Lin C Y, Chiu K C, Chang C Y, Chang S H, Guo T F, and Chen S J, Surface plasmon-enhanced and quenched two-photon excited fluorescence, *Opt. Express*, 2010, 18: 12807–12817.

In: Nanowires
Editor: Luke J. Wilson, pp. 89-129
ISBN: 978-1-63117-855-9

Chapter 3

SYNTHESIS AND PROPERTIES OF NI NANOWIRES IN POROUS SILICON TEMPLATES

S. L. Prischepa[1,*], A. L. Dolgiy[1], H. V. Bandarenka[1], V. P. Bondarenko[1], K. I. Yanushkevich[2], V. G. Bayev[3], A. A. Maximenko[3], Yu. A. Fedotova[3], A. Zarzycki[4] and Y. Zabila[4]
[1]Belarusian State University of Informatics and Radio Electronics Minsk, Belarus
[2]Scientific Practical Materials Research Center, Institute of Semiconductor and Solid State Physics National Academy of Sciences, Minsk, Belarus
[3]National Scientific and Educational Centre of Particle and High Energy Physics, Belarusian State University, Minsk, Belarus
[4]The Henryk Niewodniczanski Institute of Nuclear Physics Polish Academy of Sciences, Crakow, Poland

Abstract

Structural and magnetic properties of Ni nanowires (NWs) electrochemically deposited into pores of porous silicon (PS) template under the stationary galvanostatic regime have been investigated. Samples have

*E-mail address: prischepa@bsuir.by

been studied in detail by using scanning electron microscopy (SEM), X-ray diffraction (XRD), Raman spectroscopy and magnetic measurements. SEM analysis revealed the formation of PS/Ni nanocomposite at the initial stages of Ni deposition with the characteristic dimensions of Ni nanoparticles varying in the range 40 – 60 nm. After 60 min of deposition Ni continuous nanowires of 10 μm length have been formed. XRD analysis confirmed polycrystalline structure of Ni in the PS template with the preferential orientation along $<111>$ axis. Also some amount of Ni_2Si silicide was formed, which diffraction peak at $2\Theta \approx 33^\circ$ was especially pronounced for the small deposition times. The possible mechanism of nickel silicide formation during the electrochemical process has been discussed. It has been supposed that the presence of amorphous silicon on the pore walls facilitates the diffusion of Ni inside silicon matrix with subsequent nickel silicide formation without heating. The idea has been confirmed by the fact that on crystalline silicon the formation of nickel silicide was not observed. The magnetic properties have been investigated by studying the temperature dependence (77 – 700 K) of the specific magnetization σ. The measured σ values were lower with respect to that of bulk Ni. The effect has been explained by the influence of uncontrolled formation of nickel silicide, which causes, after heating, larger irreversibility of $\sigma(T)$ curves for samples with less deposition time. The obtained $\sigma(T)$ dependencies allowed us to determine the Curie temperature, T_C, which for small deposition times of Ni was lower (575 K) with respect to the bulk Ni (630 K). This is caused by the influence of dimensional effects on T_C value. The performed ferromagnetic resonance measurements for arrays of Ni NWs in PS templates with different porosities P allowed obtaining the crossover of magnetic anisotropy type from "easy axis" for $P< 0.7$ to "easy plane" for $P>0.7$. This crossover is induced by the growth of dipole-dipole interaction between closely packed Ni NWs with P growth. The magnetization M versus magnetic field H measurements performed for samples with $P = 0.7$ revealed the dominant role of the shape anisotropy of Ni NWs. It was confirmed both by the analysis of the squareness of the $M(H)$ hysteresis loops and anisotropy field obtained within the random anisotropy model for the magnetic field oriented either parallel or perpendicular to the NW axis.

PACS: 61.05.cp, 75.30.Gw, 75.75.Cd, 81.16.Dn.

Keywords: porous silicon, Ni nanowires, electrodeposition, specific magnetization, ferromagnetic resonance, $M(H)$ hysteresis loop, magnetic anisotropy.

1. Introduction

Nanowires (NWs) of ferromagnetic metals incorporated into porous templates of nonmagnetic materials is the subject of intensive research during the last years. Due to unique combination of the structural, electrical and magnetic properties of arrays of ferromagnetic NWs they are considered as very promising objects for magnetoelectronics [1, 2, 3, 4], magnetic recording media [5], microwave electronics [6], biological and biomedical applications [7, 8], in particular, for cancer tumor treatment [9]. Since the magnetic properties are expected to change significantly when the size at least in one direction is reduced below the single domain scale, the controllable procedure of NWs growth is of great importance. By present time different methods have been explored to fabricate ferromagnetic NWs, such as electron beam lithography [10], injection chemical vapor deposition [11], vacuum deposition [12], immersion plating [13], electrochemical deposition into different porous templates like porous anodic alumina [14, 15], polycarbonate track-etched membranes [16, 17], porous silica templates [18], nanochannel glasses [19] and porous silicon [20]. Among all these methods pore filling of porous templates by using the electrochemical deposition seems to be very attractive due to high controllability of the process, very good reproducibility and large filling factor F_f. (The F_f value is the ratio of the metal mass effectively deposited into the porous layer to the metal mass, which can be theoretically introduced into the pore channels of the template to provide a complete filling with metal.) The electrochemical method allows depositing ferromagnetic structures of different morphologies and dimensions, covering only pore walls, thus creating nanocomposite [21] or nanocylinders [22], and even growing continuous NWs with high aspect ratio [20].

Among porous templates porous silicon (PS) provides additional unique features since the pore diameters and porosity of the template can be varied in a wider range than in the cases of porous anodic alumina, polymer and glass membranes. (Here the porosity P means a fraction of the volume of voids over the total volume of the sample with thickness equal to the pore length.) In addition, the fabrication of PS and its filling with metal are in good correspondence with the modern micro- and nanoelectronic technology requirements.

PS was accidentally discovered in 1956 by Artur Uhlir from Bell Laboratory while investigating the process of silicon electropolishing in the HF-based solution [23]. Looking back in the past, we wonder now how much time has passed from the moment of discovery of PS till the idea of filling its channels with other materials, in other words, using PS as a template of exploiting the structural features of the porous layers. Studies on the electrochemical deposition of metals in the PS were started in the middle of the 80's of the last century. The main driving force behind this research was the idea of getting thick layers of silicide by filling the channels of PS with metal followed by thermal annealing. Silvia Tsao from the Texas Instruments first applied in 1986 the CVD method to fill pores of PS with tungsten to form thick layer of tungsten silicide [24]. The first work on the electrochemical deposition of nickel into PS was published in 1985 [25]. The electrochemical technique was also used to deposit Co [26], Fe [27], Er [28, 29] in PS. Critical analysis on impregnation of the PS template with different materials is presented in [30]. The first results on the anisotropy of the magnetic properties of Ni NWs in the matrix of PS were observed in 1994 [20], while the influence of the porosity of the PS template on the type of magnetic anisotropy was revealed only 15 years later [31]. Recently the critical analyzes of the current state-of-the-art of the employment of PS as the template for different magnetic nanocomposites have been presented [22, 32, 33, 34].

Fabrication of PS/metal nanocomposites comprises two main stages: (a) obtaining the PS matrix with desired pore diameter and length, and (b) filling the pore channels with metal. The situation of completely filled pores when metal NWs inheriting pore sizes are formed is of the main interest. Still the partial filling process, when on the surface and on the pore walls metal nanoparticles (NPs) are formed with sizes less than the pore diameter, is attractive as well.

Actually, metal deposition inside pores of PS is still nontrivial task. Formation of metal NWs with the aspect ratio greater than 100 and the filling factor higher than 0.25 is still challenging [33]. The serious problem is related to the exchange of reagents within the pore, which influences the F_f value. Morphological changes of the pore walls like dendritic branches as well as hydrogen evolution lead to inhomogeneity of the metal deposit so it is very difficult to obtain homogeneous filling of the pores. Recently it was demonstrated that the application of very high magnetic field ($B = 8$ T) during the anodization of the PS template allows reducing the dendritic structure of the pore walls thus creating the net interface between the pore walls and NW [35]. But this method requires liquid helium temperatures for superconducting solenoid and can not be considered for commercial technology.

If the exchange of reagents within the pore is limited, the channel might be blocked by the accumulation of the deposited metal. This prevents uniform filling of the pores along its entire length. One possible way to overcome this problem is a pulsed electrochemical deposition of metal into PS. At the present time, use of the pulse mode allows forming Ni NWs in the matrix of PS with the diameters ranging from 20 to 100 nm and length of several micrometers [20, 32, 33].

In this chapter results of the study of the Ni deposition into pores of PS under the *stationary* galvanostatic regime from a modified Watts bath is presented. We fabricated PS samples of different porosities $P = 0.5, 0.7$ and 0.8 with pore length up to 10 μm and diameter of pores varying in the range $60 - 80$ nm. Due to the peculiar developed process of the electrochemical deposition, pores were well filled with Ni without use of the pulsed mode. The exhaustive investigation of the structural and magnetic characteristics of arrays of Ni NWs revealed the influence of both the PS template and geometry of NWs on their magnetic properties.

2. Synthesis

2.1. Porous Silicon Templates

In this work PS was formed by anodization of single crystalline silicon in HF-based solution. Actually, producing PS with required pore dimensions and thickness is a specific process. Finding ways to improve the process of PS anodizing is still an important challenge for researches. Several process parameters play the key role in the technology of PS.

Type of starting silicon wafers has a determinant influence on the structure of PS formed by anodization. To ensure reproducible parameters of the porous matrix it is better to use the highly doped p+ type (i.e. boron doped) and n+ type (i.e. antimony doped) wafers with the resistivity ranging from 0.01 to 0.005 $\Omega\times$cm. For these low resistivity wafers there is no need in the metal film on the backside of the wafer to ensure uniform current flow. PS formed on such highly doped wafers possesses good thickness and porosity uniformity over the wafer diameter and high reproducibility of PS parameters from sample to sample.

The concentration of HF in the electrolyte is the second important parameter, which influences the structural parameters of PS. For the anodization of highly doped silicon wafers aqueous or mixed aqueous-alcoholic solutions are used with the concentration of HF varying from 10 to 32%. In case of aqueous solutions a rigorous stirring of the electrolyte is required to ensure the removal of hydrogen bubbles evolved during the dissolution of silicon from the surface of PS. Otherwise non-uniform PS layers may be formed.

Electrical anodization regime is the third major technological factor that affects the kinetic of formation and structural parameters of PS. Typically galvanostatic mode is used when the current is maintained constant by a galvanostat as current source, thus allowing a thorough control over the PS thickness. For electrolytes containing 10 – 32% HF, the recommended anode current density (j_a) lays in the range of 20 – 150 mA/cm^2.

In the present work antimony doped 100 mm single crystalline silicon wafers with (100) orientation and 0.01 $\Omega\times$cm resistivity were used as the initial substrates [36, 37]. Silicon wafers were cleaned by using the standard RCA solution, dried in the centrifuge and cut into a number of rectangular 9 cm^2 samples. Just before PS formation the experimental samples were immersed into 5% HF solution to remove the native oxide. Immediately after that, the Si samples were disposed in a standard electrolytic cell made of Teflon. Uniform PS layers were formed by an electrochemical anodization of silicon samples in 1:3:1 =HF:H_2O:$(CH_3)_2CHOH$ solution. The anodized samples were rinsed in the deionized water for 5 min. A spectrally pure graphite disk was used as a contact electrode to the back side of the samples during the electrochemical treatment. Platinum spiral wire was used as a cathode electrode. Anodization was carried out at a current densities j_a = 10, 80 and 92 mA/cm^2 for 200 s. These regimes provided formation of uniform PS layers with thickness of 10 μm and porosity of 0.5, 0.7 and 0.8, respectively [38]. Pore diameters of 60 – 80 nm were estimated from the SEM images. In Figure 1 we present the typical top (a) and cross sectional views of the PS template with $P = 0.7$.

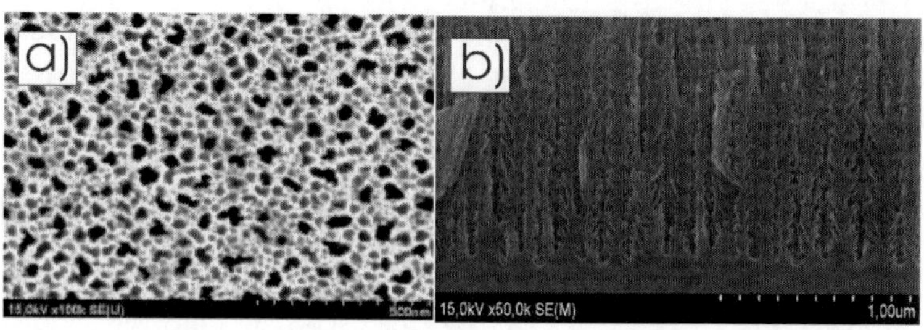

Figure 1. SEM images of the top (a) and cross section (b) of the PS template with $P = 0.7$.

2.2. Ni Electrochemical Deposition into PS Templates

2.2.1. Electrochemical Process

After PS formation the HF solution was removed and the electrolytic cell was thoroughly rinsed with deionized water for 5 min. The cell was then filled with the electrolyte for the deposition of nickel. Nickel was incorporated into PS template by electrochemical galvanostatic deposition at constant current density of 3.5 mA/cm^2. The composition of electrolyte for Ni deposition was slightly different from that known for Watts bath. It consisted of 213 g/l $NiSO_4 \times 7H_2O$, 5 g/l $NiCl_2 \times 6H_2O$, 25 g/l H_3BO_3 and 3 g/l saccharin. At room temperature such electrolyte has $pH = 2.6$. Ni was deposited into PS templates for different times in the range 30 s$-$80 min. For the sake of clarity, the samples were named using the initials PS (porous silicon) or CS (crystalline silicon) followed by N (for nickel) and a number indicating the nominal time of Ni deposition, t_{Ni}, in minutes. For example, PSN60 is for Ni deposited in PS template for $t_{Ni} = 60$ min, while CSN0.5 is for Ni deposited onto c-Si substrate for $t_{Ni} = 30$ sec. Thickness of the porous layer always was 10 μm. Potential measurements were carried out using Ag/AgCl reference electrode filled with the saturated KCl solution. The reference electrode was immersed into a small bath filled with the solution for Ni deposition. This bath was connected with the electrolytic cell by a flexible polymer tube of 2 mm inner diameter ended with a Luggin glass capillary of 200 μm aperture. Both the Luggin capillary and polymer tube were filled with the solution for Ni deposition. The Luggin capillary was placed on the surface of PS and defined a clear small sensing point for the reference electrode near the PS electrode. The equipment used to conduct electrochemical processes was the potentiostat/galvanostat AUTOLAB PGSTAT302. Gravimetric method was applied to determine the porosity of PS and the filling factor of Ni. Mass measurements were performed with Sartorius CP225D micro/analytical electronic balance. The instrumental error of mass measurements was about 10 μg.

The structure of the samples was studied with the scanning electron microscope (SEM) Hitachi S-4800. The elemental composition of samples was determined using Cambridge Instruments Stereoscan-360 with a Link Analytical AN 10000 energy dispersive X-ray analyzer. The crystal structure was studied by $\Theta - 2\Theta$ X-ray diffraction (XRD) using a modified DRON-3M diffractometer with CuKα ($\lambda = 0.154242$ nm) radiation (Bourevestnik Inc., St. Petersbourg, Russia). Measurements were performed at room temperature with the step in angle $\Delta 2\Theta = 0.03^o$. Acquisition time for each point was $\Delta t = 3$ s. Some peaks were acquired with the precision $\Delta 2\Theta = 0.01^o$.

Nanofinder HE (Lotis-TII) confocal spectroscopy instrument was used to perform the Raman investigation. The Raman spectra were taken with the spectral resolution of 3 cm^{-1} in the backscattering geometry at ambient conditions using a 532 nm wavelength excitation laser.

2.2.2. Kinetics of Pore Filling with Ni

One of the important tasks of the electrochemical deposition of metals into PS is the determination of the distinctive time points, which correspond to different filling stages. The moments of the initial precipitation of metal on the pore walls and the complete filling of the whole pore volume are especially important. To solve such problem the applied electrochemistry offers measurement of the potential of the PS surface during the deposition in galvanostatic mode. The measurement of PS potential as a function of deposition time may provide considerable insight on the different deposition behaviors [13]. Figure 2 shows the dependencies of the PS potential on Ni deposition time for five PS samples with different t_{Ni}. Data are for the PS template with $P = 0.7$.

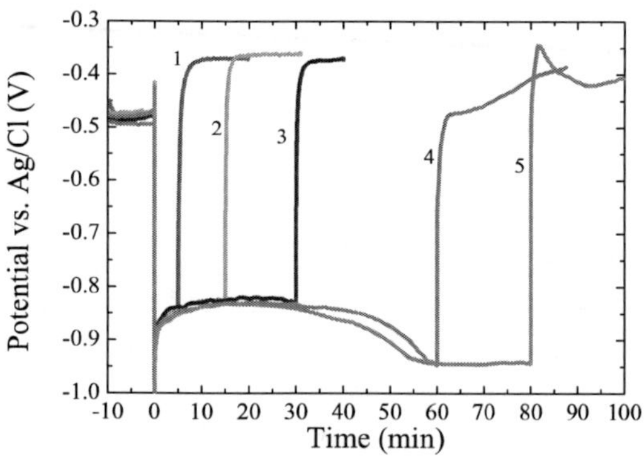

Figure 2. The potentials of PS versus time of nickel electrochemical deposition. Deposition process was stopped at different times: (1) 5 min, (2) 15 min, (3) 30 min, (4) 60 min and (5) 80 min, respectively.

Measurement of the PS potential was started immediately after filling the cell with the electrolyte and the cathod current was applied 10 minutes later. In this set of samples Ni was deposited in PS during different periods of time: 5, 15, 30, 60 and 80 minutes, as it is marked in Figure 2. After stopping the deposition (these points correspond to the decreasing spikes of the potential) the potential was recorded a while longer. The potential value in the electrolyte without external polarization was almost constant in time and fluctuated around $-0.475\ldots-0.485$ V. It indicates no appreciable change in reactions on PS surface, i.e. the surface of PS was stable in the Watts solution and did not change [13]. At the moment of the current turn-on the sharp increase of PS potential up to $-0.975\ldots-0.98$ V was observed. Afterwards the potential was gradually reduced and in 12 min reached permanent value of about $-0.83\ldots-0.84$ V. Half an hour later PS potential started again slow rising which stopped at $55-60$ min at the value of -0.95 V. Further deposition did not change this value. Current turn-off at 5, 15 and 30 minutes (curves 1, 2 and 3) led to the immediate potential decreasing to $-0.37\ldots-0.375$ V. The same reduction of the potential was observed after current stops at 60 and 80 minutes (curves 4 and 5, respectively) but it took more than 30 minutes to reach constant value of -0.38 V [36]. Below it will be shown that such potential behavior after 60 min of the process indicates a full coverage of the PS surface with Ni.

To study the kinetics of pore filling we have used scanning electron microscopy (SEM) images of cross sections of the experimental samples after different periods of the metal deposition. As an example, Figures 3-5 presents SEM images of the top, middle and bottom parts of cross sections for the PSN5, PSN15 and PSN80 samples, respectively [37]. Also the whole cross section of the samples is shown (marked with letter (a) in Figures 3-5).

There are only a few nanoparticles (NPs) of nickel observed on the pore walls in the top area of the PSN5 sample (Figure 3b). In the central part of the PSN5 sample (Figure 3c) NPs of $30-50$ nm in diameter are seen. The average linear density of NPs is about 57 per 1 μm of pore length. The bottom image of the PSN5 sample (Figure 3d) shows the interface between PS and single crystalline Si. It is characterized by significant increase of NPs linear density to several tens per 1μm. Such fact means that the metal deposition process mostly began in the lower part of the pore channels. More careful analysis of the SEM images of the PSN5 sample (Figures 3b - 3d) reveals that the primary Ni NPs nucleation occurred on the tips of the PS skeleton branches. We suggest it was caused by high local current density due to accumulation of the electric field on the tiny ends of the Si elements of the PS skeleton. The pore entrances were

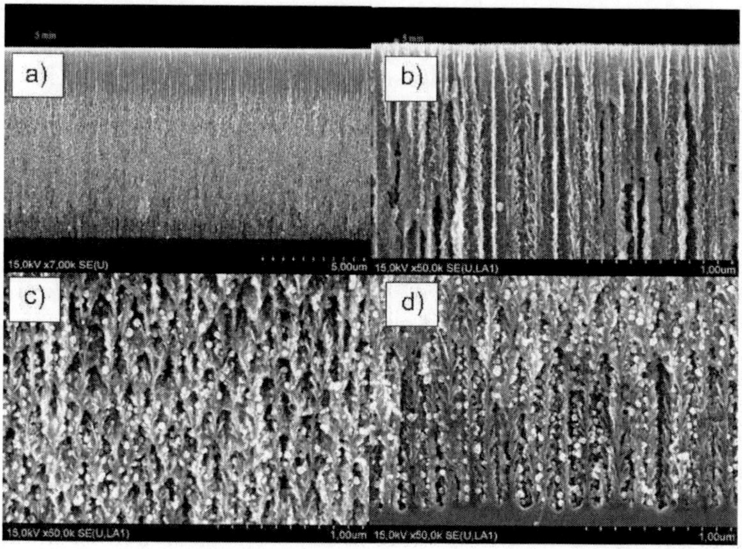

Figure 3. The cross section of the PSN5 sample. (a) the whole cross section, (b) the upper, (c) middle and (d) bottom parts.

not blocked as well as Ni deposition on the outer surface of PS did not take a place. The F_f value for the PSN5 sample was estimated by gravimetric method as 0.045.

Figure 4 shows SEM images of cross section of the PSN15 sample. The upper and middle regions of PS (Figures 4b, 4c) look like the same areas of the PSN5 sample but are characterized by relatively larger density of NPs. However, in the bottom part we see great differences. The dimensions of Ni NPs are about $40 - 70$ nm as in the center but their density is several times more. Moreover, Ni NPs coalesced and formed quasi continuous wires. The F_f value of the described sample was about 0.23.

Figure 5 presents SEM cross sections images of the PSN80 sample. This outer surface of the sample was covered with the continuous nickel film of 2 μm thickness. The upper Ni layer is not observed on Figure 5b, because it was mechanically removed before the sample's cutting for the SEM analysis. Pore channels in this case are well filled with metal NWs which represent nanopillars of the large tightly contacted conglomerates of Ni. On the top of the cross section grown together particles are observed. The diameter of the particles

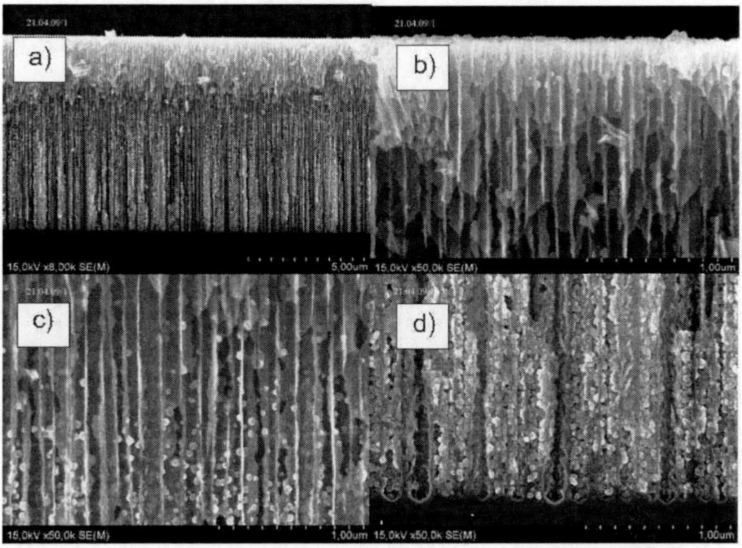

Figure 4. The cross section of the PSN15 sample. (a) the whole cross section, (b) the upper, (c) middle and (d) bottom parts.

varies in the range from 100 to 120 nm. It should be noted that pretreatment of the sample, resulting in the exfoliation of the surface layer of Ni, could destroy the Ni NWs in the upper part of the sample. In the middle of the PSN80 sample nickel NWs are rather monolithic and their diameters are about $60 - 80$ nm. It exactly corresponds to the diameters of the pore channels. At the bottom of the PSN80 sample the completeness of nickel filling is lower than in the central part. It is possible to see the individual NPs of nickel. The filling factor was estimated as $F_f \approx 0.67$. Actually, this value of the filling factor could be overestimated, because during the mechanical removal of Ni from the surface some part of the material could remain on the top of the sample [31]. This additional mass of Ni was taken into account during the F_f calculations. More deeper considerations regarding the evaluation of the filling factor will be reported in Section 4.2.

Detailed study of the kinetics of Ni deposition in the PS templates required X-ray microanalysis of the cleavage surface of the samples. The profile of distribution of the Ni concentration from the surface to the bottom of the porous layer was obtained. Results of the analysis are shown in Figure 6. We analyzed the profiles of nickel in PS (Figure 6) and data on the morphology of the samples

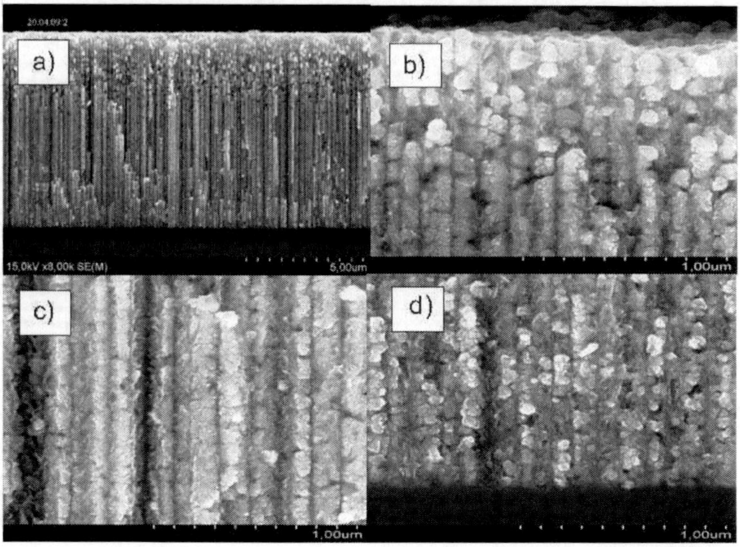

Figure 5. The cross section of the PSN80 sample. (a) the whole cross section, (b) the upper, (c) middle and (d) bottom parts.

(Figures 3 - 5) for different times of Ni deposition and proposed the following kinetics of the Ni NWs growth.

The nucleation and growth of Ni NPs begin at the initial stages (less than 5 minutes) in the lower region of PS (in the depth of $8 - 10$ μm from the surface) on the tips of branches of the PS skeleton. The diameters of NPs are around $40 - 70$ nm. Prolongation of the deposition time to 15 minutes leads to the increase of the density of nickels NPs in the central and bottom parts of PS. It is mostly caused by increase in the number of Ni NPs rather than by change of their dimensions which grow very slightly. In that way Ni NPs coalesce and almost completely fill the pore channels with deposition for 60 minutes as it is proved by constant potential value -0.95 V (Figure 2). Further deposition causes Ni continuous layer formation on the outer surface of PS (80 minutes, Figure 5a). The consideration of the obtained results allows summarizing the PS potential stabilization after 60 minutes of the Ni electrochemical deposition is related to the emergence of Ni from the porous layer on the outer surface of PS. In that way control of the PS filling level during the formation of Ni NWs in the PS template might be performed [37].

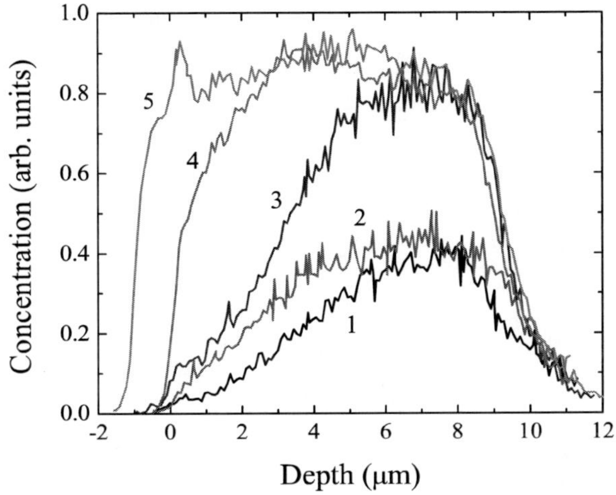

Figure 6. Nickel concentration profiles in PS samples determined by X-ray microanalysis. Nickel deposition time: (1) 5 min, (2) 15 min, (3) 30 min, (4) 60 min and (5) 80 min, respectively.

3. Crystalline Structure of Ni Deposited in PS

XRD analysis was carried out to investigate the crystalline structure of Ni deposited in the PS templates. It is known that crystalline nickel has face centered cubic (fcc) cell with the lattice parameter $a = 0.3524$ nm [39].

In Figure 7 we present XRD patterns acquired for the samples with different times of Ni deposition, PSN0.5 (Figure 7a), PSN1 (Figure 7b), PSN5 (Figure 7c) and PSN60 (Figure 7d). Analysis of the Figure 7a shows that after 30 s of deposition the non-ferromagnetic metallic Ni_2Si (111) phase is dominant. For the PSN1 sample the amplitude of Ni_2Si (111) peak increases which reflects the increase of the total amount of this silicide. At the same time the peak of pure Ni(111) appears at $2\Theta \approx 44.5^o$, which means that at the very initial stages of electrochemical deposition ($t_{Ni} \leq 30$ s) the nickel silicide Ni_2Si is only formed. The pure Ni starts to deposit only after the 1st minute of the process. At that the polycrystalline Ni is deposited with orientations (111), (200) and (220), Figure 7c. These orientations are typical for electrodeposited Ni nanowires [40]. With the subsequent increasing of the deposition time we observed that the intensity of Ni_2Si peak is decreased while the amplitude of Ni peaks is increased. At the long deposition time, $t_{Ni} = 60$ min, the Ni_2Si (111) peak is completely vanishing. Nevertheless, the observed broadening of the Ni peaks could be caused by the presence of nickel silicide even at high t_{Ni} values, as indicated in Figure 7c.

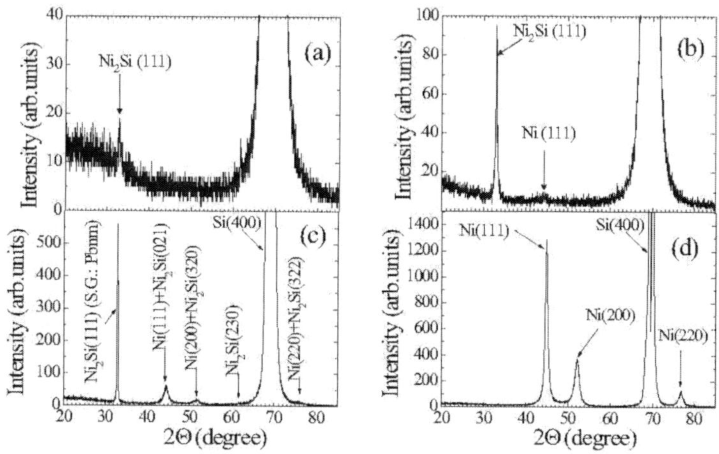

Figure 7. XRD of PS subjected to the Ni electrochemical deposition. (a) PSN0.5 sample, (b) PSN1 sample, (c) PSN5 sample and (d) PSN60 sample.

The observed formation of silicide is not surprising. High temperature annealing of metal-silicon hybrids normally results in the formation of silicides [41]. For the Ni-Si case nickel diffuse into the Si lattice resulting in a large amount of metal nonferromagnetic phases [41]. Generally, such diffusion process is stimulated by the grain boundaries, surface and bulk defects, temperature [42]. Annealing of thin Ni films deposited on Si leads to gradual formation of the three dominant Ni-Si phases. Firstly Ni_2Si is formed at $T \approx 200^o$ C [43], which with further increase of temperature up to 400° C is converted to the low resistive and less strained monosilicide NiSi. Annealing at higher temperatures leads to the progressive consumption of NiSi by disilicide $NiSi_2$ that fully completes at $T \approx 800^o$ C [41] Nevertheless, at the initial stages of Ni deposition on Si substrate, the Ni_2Si disordered islands could be formed even at room temperature [44]. This occurs because the energy released by metal cluster formation can be sufficient to overcome the activation barrier for the diffusion [45]. The room temperature formation of Ni_2Si can be also conditioned by ion beam mixing process [46, 47].

To clarify the mechanism of Ni_2Si formation and the possible role of porous structure on it we deliberately deposited Ni onto a single crystal Si(100) wafers (c-Si). Two samples were prepared, with the deposition times $t_{Ni} = 30$ s (sample CSN0.5) and $t_{Ni} = 60$ s (sample CSN1). The XRD patterns for these samples are shown in Figure 8. It is clearly seen that there are no traces of Ni_2Si in this case. For CSN0.5 the small amount of Ni (111) presents, Figure 8a, while for CSN1 both Ni (111) and Ni (200) peaks present, the strength at (111) exceeds that at (200), as for Ni nanowires.

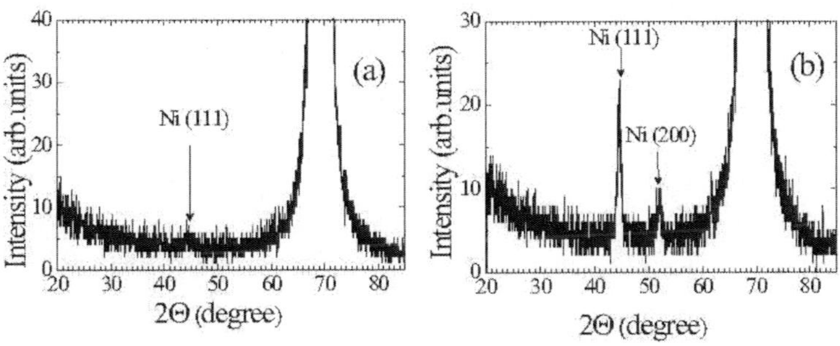

Figure 8. XRD patterns of c-Si with electrochemically deposited Ni. (a) CSN0.5 sample and (b) CSN1 sample.

From this result one may conclude that the presence of pores plays an important role during the formation of nickel silicide at ambient temperature. We propose that the nickel silicide formation at room temperature during the electrochemical process is caused by the specific structure and composition of pore walls. Indeed, PS is partially consisted of amorphous silicon (a-Si) with a lot of broken bonds [48]. The presence of a-Si in our samples was detected directly by means of Raman spectroscopy. In Figure 9 we show the Raman spectra of the cross section of PS (solid line) and c-Si (dashed line). The laser spot was focused on the several places of each sample to obtain average information. The example of the optical image of laser spot on the cross section of PS template is shown in Figure 10.

Crystalline silicon has highly uniform bond angles and bond lengths and exists in a limited number of states. This results in sharp Raman peaks with a characteristic strong band at 521 cm^{-1} [49]. This sharp peak is seen in Figure

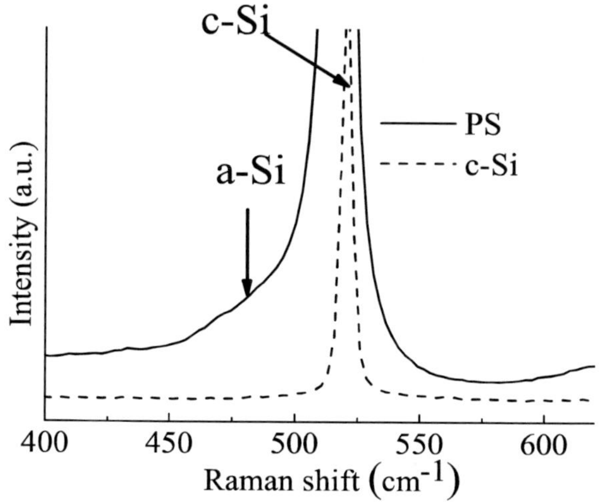

Figure 9. Raman spectra of the cross section of PS (solid line) and c-Si (dashed line). Arrows indicate characteristic positions of the Raman bands for c-Si (521 cm^{-1}) and a-Si (480 cm^{-1}).

9 for the c-Si sample. Amorphous silicon is less orderly in its arrangement with a wider array of bond angles, bond energies and bond lengths in addition to dangling bonds. The distribution of possible states leads to a broad Raman band centered at about 480 cm^{-1} that is readily distinguishable from that of crystalline silicon [49]. This broadening around Raman shift for a-Si is clearly seen in Figure 9 for the PS sample.

It is known that Ni ions relatively easy diffuse into Si through its defects [41]. It is reasonably to suppose that an external electric field provides Ni ion flow to PS surface and further into PS itself. Moreover, the activation energy for the growth of the Ni_2Si phase at the Ni/a-Si interface is much less (4.48×10^{-20} J [50]) with respect to the Ni/c-Si one (2.4×10^{-19} J [51]). In this way the a-Si surface of PS becomes filled with Ni ions, but it is not yet silicide, because Ni_2Si formation requires an extra energy which is usually provided by annealing [41]. In our case a-Si enriched with Ni ions converts to Ni_2Si silicide most likely due to the suitable energetic conditions such as external electric field and contact potential at the solution/PS interface. The observed decomposition of the Ni_2Si phase at large t_{Ni} could be caused by the peculiarities of the electro-

Figure 10. The optical image of laser spot on the cross section of the PS template during the acquisition of a Raman spectrum. The white scale bar is 0.5 mm.

chemical processes at the solution/PS interface [21]. Indeed, usually Ni_2Si on Si is stressed and these stresses are developed during the silicide growth [52]. Thus silicide formation stops when the energy of the electrochemical system is not enough to overcome the formation energy barrier [52]. On the other hand Ni_2Si is the material of metallic conductivity [41] and PS/Ni_2Si junction field promotes an electron extraction from the Si skeleton. So, stressed Ni_2Si could begin to relax and fill with electrons. In this case Ni ions in silicide layer reduce to atomic nickel as do their counterparts from the solution/PS interface. In that way Ni_2Si decomposition could be performed.

4. Magnetic Properties of Arrays of Ni NWs in PS Templates

4.1. Specific Magnetization and Curie Temperature

All the synthesized samples with $t_{Ni} \geq 5$ min possess a ferromagnetic order. This was confirmed by measuring the specific magnetization σ at different temperatures T, ferromagnetic resonance (FMR) at room temperature and isothermal magnetic hysteresis loops $M(H)$ at different T.

For the measurements of the specific magnetization the equipment based on the ponderomotive method was used [53]. The ponderomotive method consists of measurements of the mechanical (ponderomotive) force acting on sample placed in the magnetic field. The equipment [53] allows measuring both the magnetization and the magnetic susceptibility as functions of temperature in the range 77 – 1400 K using small, with diameter less that 5 mm, samples, that minimizes the temperature gradient across them. In this study the ponderomotive force was measured by Sartorius 310S micro-analytical electronic balance with the precision of $\pm 5 \times 10^{-5}$ N. The applied magnetic field was 860 mT, which is well above the saturation field H_{sat} for Ni ($H_{sat} \approx 300$ mT, see Section 4.3). Therefore, the specific magnetization values in this case correspond to the saturation magnetization M_{sat}. The precision of measurements was ± 0.01 A$\times$m$^2\times$kg^{-1}. To avoid the oxidation during heating the samples were put in a vacuum ampoule under the pressure 1.33×10^{-3} Pa.

For the $\sigma(T)$ measurements samples were cooled down to liquid nitrogen temperature (77 K), then the magnetic field of 860 mT was applied and the static ponderomotive force was measured at different temperatures up to 700 K. After that measurements were continued by cooling samples down to room or liquid nitrogen temperatures. The specific magnetization values σ were extracted by knowing the ponderomotive force and the mass of deposited Ni [37]. In Figure 11 the measured dependencies $\sigma(T)$ are presented for the samples with $t_{Ni} = 5$ min (Figure 11a), 15 min (Figure 11b), 30 min (Figure 11c) and 60 min (Figure 11d).

The analysis of the results presented in Figure 11 revealed that the absolute values of the low temperature specific magnetization of Ni grown in the PS template are always less with respect to low temperature specific magnetization known for the Ni powder ($\sigma_{powder} \approx 58.9$ A$\times$m$^2\times$kg $^{-1}$) [54]. These less σ values are naturally explained by the presence of nickel silicide in the samples. In this case the loss of electrons of Ni while creating the Ni-Si bonds is especially sensitive to the magnetic properties. Some discrepancy in σ values for different samples could be related to the uncertainty in the Ni mass determination and, which is more important, to uncontrollable nickel silicide formation in our samples, including possible amorphous phases [50].

From the results presented in Figure 11 it follows also that, after heating to 700 K, the temperature dependence of σ becomes irreversible. Indeed, the $\sigma(T)$ curves, measured in cooling regime, reveal always less σ values. Moreover, for the PSN5 sample (Figure 11a), the cooling $\sigma(T)$ curve shows the diamagnetic behavior caused by the Si substrate. It means that in this sample Ni was completely converted to nonmagnetic silicides during the heating in vacuum. For the PSN15 sample (Figure 11b) we observed the paramagnetic behavior after heating. For the PSN30 (Figure 11c) and PSN60 (Figure 11d) samples, which cor-

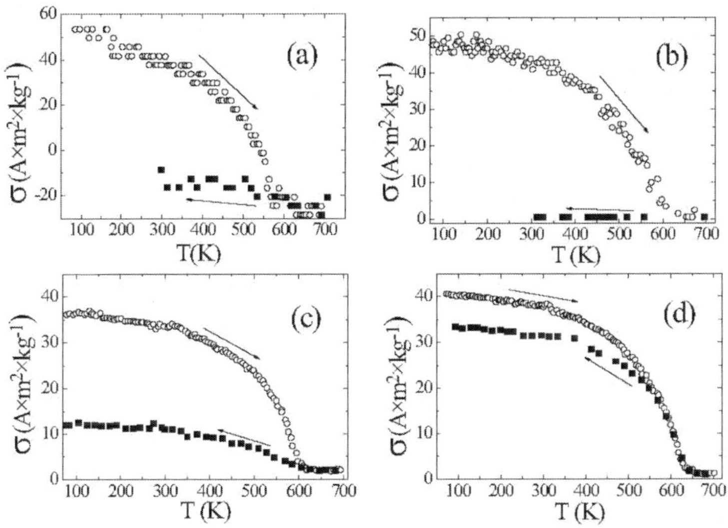

Figure 11. Temperature dependence of the specific magnetization σ for the PS/Ni samples formed at different times of Ni deposition. (a) PSN5, (b) PSN15, (c) PSN30 and (d) PSN60 samples, respectively.

respond to Ni NWs, the cooling behavior becomes strongly ferromagnetic. At that the irreversibility of $\sigma(T)$ curves measured in heating and cooling regimes becomes lower reflecting the less influence of nickel silicide formation on the magnetic properties for the larger mass of deposited Ni.

From the results presented in Figure 11 it is possible to extract the Curie temperature, T_C, for arrays of Ni NWs. Indeed, according to the Weiss theory, the specific magnetization close to T_C depends on temperature as $\sigma(T) \sim (1 - T/T_C)^{1/2}$. In the Figure 12 we plot the σ^2 versus T dependence for the PSN15 (Figure 12a) and PSN60 (Figure 12b) samples, which gives the possibility to evaluate the T_C value. From that we obtained $T_C = 592$ K for the PSN15 and $T_C = 624$ K for PSN60 (see arrows in Figure 12) samples. These values are slightly lower than that known for bulk Ni, $T_C = 627$ K [55]. The same data elaboration for T_C evaluation was performed for each sample. As a result, in the Figure 13 we show the behavior of T_C versus t_{Ni}. The dashed line corresponds to the Curie temperature of bulk Ni. It is seen that, while increasing the deposition time, the T_C values approach that of bulk material. The main reason of less measured T_C values in our samples is related to the size effect. Indeed, PSN5 sample consists of Ni NPs of $40-60$ nm in diameter embedded in PS template, see Figure 3. The same morphology was observed for the PSN15 sample, Figure 4. The Curie temperature of such small nanoparticles usually is less with respect to bulk values due to size effects [56]. The PSN60 sample consists of continuous closely packed Ni NWs, see Figure 5. As a result, the T_C for this sample is almost the same as for bulk material.

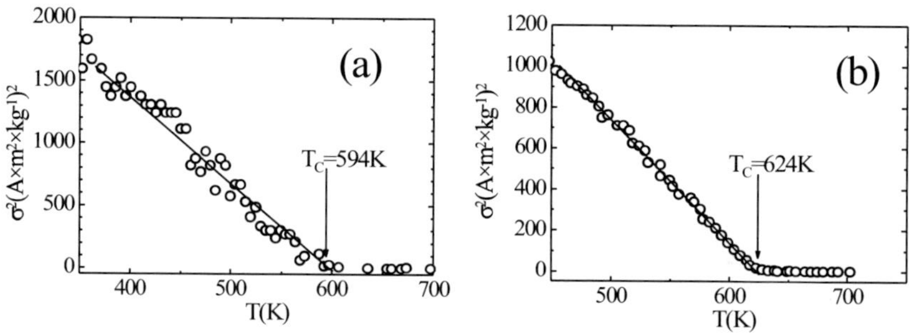

Figure 12. σ^2 versus T dependence. Lines represent the best fit to the experimental data. Arrows indicate the evaluated Curie temperatures. (a) PSN15, (b) PSN60 samples, respectively.

4.2. Ferromagnetic Resonance in Arrays of Ni NWs

Ferromagnetic resonance (FMR) is one of the varieties of electron magnetic resonance, revealed as the selective absorption of ferromagnetic electromagnetic energy at frequencies coinciding with the natural frequencies of precession of the magnetic moments of the electron system of the ferromagnetic sample in an internal effective magnetic field. Uniform FMR can be detected by the magnetic radio spectroscopy. As the internal effective magnetic field of the sample is specified by a magnetic anisotropy the method of FMR spectroscopy serves as a useful technique for evaluating the magnetic anisotropy of Ni NWs. The investigation of the FMR in arrays of Ni NWs embedded in PS templates gives the opportunity to determine the magnetic anisotropy peculiarities of the whole system and single Ni NW specifically.

For the FMR study the spectrometer of electron paramagnetic resonance

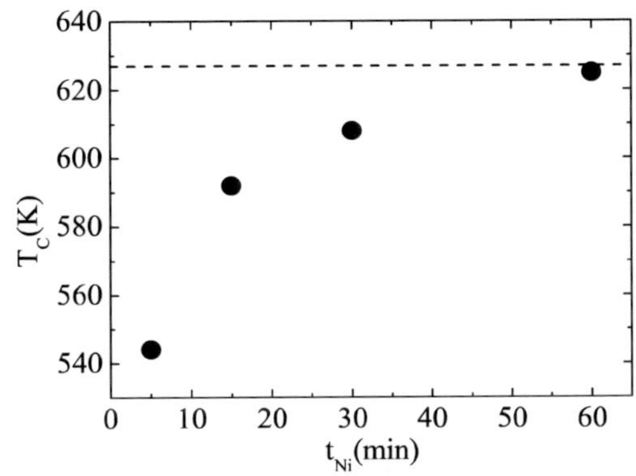

Figure 13. Curie temperature T_C versus t_{Ni} for PS/Ni samples. Dashed line corresponds to the value of T_C for bulk Ni.

(EPR) "Radio PAN Se/X-2543" in $X-$ band on the frequency $f = 9.32$ GHz was used. The polarized alternating magnetic field with the amplitude of modulation $\Delta B = 0.1$ mT and the frequency $f = 100$ kHz was applied to the sample. As a result, the first derivative of the absorption line was registered, that increases the signal to noise ratio [57]. Measurements of the value of the polarizing field were performed by the nuclear magnetic resonance magnetometer and Hall sensor, the frequency was measured with the wavemeter. To control the quality of the H_{102} resonator and tune the phase of registration of the magnetic resonance signal of the ruby crystal (R_y) located on the wall of the resonator was used. The investigated samples were located at the resonator center. The measurements were performed at room temperature at dark in air. The orientation of the samples in the external magnetic field was characterized by the polar angle θ_B between the magnetic induction $\mathbf{B}$ and normal to surface of the sample $\mathbf{n}_z$, see Figure 14a. Therefore, when the magnetic field was applied perpendicular to the NW axis, i.e. parallel to the surface substrate, it corresponds to $\theta_B = 90^o$ (in plane (IP) configuration, see Figure 14b). The magnetic field orientation parallel to the NW axis means $\theta_B = 0^o$, which corresponds to the out of plane (OOP) configuration, see Figure 14c.

Spectra of the magnetic resonance for the sample with the porosity $P = 0.7$ for OOP ($\theta_B = 0^o$) and IP ($\theta_B = 90^o$) orientations of the magnetic field are shown in Figure 15. The observed weak signal at $B = 151.7$ mT represents

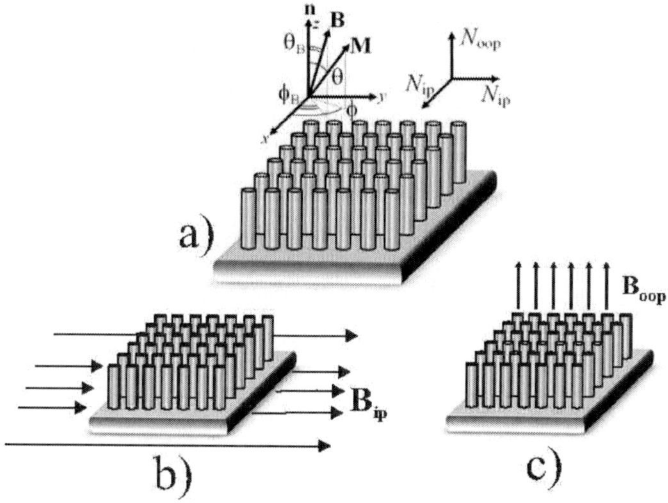

Figure 14. Schematic representation of the arrays of Ni NWs in the external magnetic field of different orientation. (a) The magnetic moment **M** in the tilted external magnetic field **B**. $\mathbf{n}_z$ is a normal to the sample surface. θ and ϕ are angles of the spherical system of coordinates between **M** and $\mathbf{n}_z$, respectively. θ_B and ϕ_B are angles of the spherical system of coordinates between **B** and $\mathbf{n}_z$, respectively. N_{ip} and N_{oop} are demagnetizing factors in cases of in plane and out of plane orientations, respectively. (b) The IP configuration of the external magnetic field for the arrays of Ni NWs. (c) The OOP configuration of the external magnetic field for the arrays of Ni NWs.

the EPR signal from a quartz sample holder. Lines designated as R_y, are EPR signals of paramagnetic Cr^{3+} ions in the ruby crystal placed on the wall of the cavity [58]. As it can be seen from the Figure 15, the signal of the first derivative of the absorption of the microwave field by the sample is intense broad line. Moreover, the value of the resonant magnetic induction B_r depends on the orientation of the magnetic field with respect to the NW axis, which indicates the presence of the ferromagnetic resonance. The peak-to-peak widths of the measured curves are $\Delta B_{pp} = 110$ mT at $\theta_B = 0°$ and $\Delta B_{pp} = 135$ mT at $\theta_B = 90°$. FMR spectra were obtained at different θ_B values. The original FMR spectra in the range $0° \leq \theta_B \leq 90°$ are shown in the Figure 16.

On the base of the measured FMR spectra it is possible to plot the dependence of the resonant field on the angle θ_B. This result is shown in the Figure 17b. The obtained angular dependence of B_r indicates the presence of the easy axis magnetic anisotropy, which coincides with the axes of Ni NWs. In the Figure 17 we also show the obtained $B_r(\theta_B)$ dependences for arrays of Ni NWs in PS with $P = 0.5$ (Figure 17a) and $P = 0.8$ (Figure 17c). These results will be discussed later.

From the $B_r(\theta)$ dependence such important micromagnetic parameters as the spectroscopic splitting factor g (g-ratio) and effective anisotropy filed could be assessed. To do this we assume that pores in PS are filled uniformly by Ni and the sample is magnetized homogeneously by the external magnetic field. In the case of the classical recording of a FMR spectra the frequency of the microwave field is constant, and the external magnetic field varies linearly with time. In this case the resonant frequency is determined by the following dispersion relation [59]:

$$\left(\frac{\omega}{\gamma}\right)^2 = \frac{E_{\theta\theta}E_{\phi\phi} - E_{\theta\phi}^2}{(M_{sat} \times sin\theta)^2}|_{\theta=\theta_0,\phi=\phi_0}, \quad (1)$$

where $\omega = 2\pi f$, f is a frequency of the microwave field, $\gamma = g\mu_B/\hbar$ is gyromagnetic ratio, μ_B is Bohr magneton, $E_{\theta\theta}$, $E_{\phi\phi}$ and $E_{\theta\phi}$ are the second derivatives of the free energy E with respect to the angles (θ, ϕ) of the spherical system of coordinates (see Figure 14) calculated at angles (θ_0, ϕ_0) which correspond to the equilibrium magnetization. The latter is determined by the condition of the minimum of the free energy, $E_\theta = 0$, $E_\phi = 0$.

Free energy for the homogeneously magnetized system in the external magnetic field B is expressed as [60]

$$E = -\frac{\mathbf{M}\,\mathbf{B}_{eff}}{\mu_0}, \quad (2)$$

where $\mathbf{M}$ is a vector of the magnetization of the sample, $\mathbf{B}_{eff}$ is a vector of the induction of the resulting magnetic field inside a sample, which is a result of superposition of the external magnetic field B and internal (induced) field B_{ind}, μ_0 is the magnetic constant.

Actually the induced magnetic inductance B_{ind} depends on the field of the magnetocrystalline anisotropy B_{cryst} and demagnetized field B_{demagn}. The latter is proportional to the magnetization of the sample [61]

$$\mathbf{B}_{demagn} = -\mathbf{NM}, \quad (3)$$

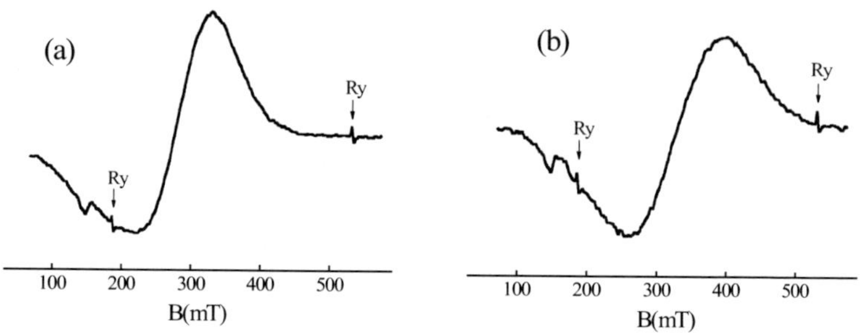

Figure 15. FMR spectra of the array of Ni NWs in PS with $P = 0.7$. R_y is the electronic paramagnetic signal of ruby crystal. (a) OOP configuration ($\theta_B = 0^o$), (b) IP configuration ($\theta_B = 90^o$).

where **N** is a demagnetizing tensor, which characterizes the magnetic anisotropy of the sample. The XRD study proved the polycrystalline nature of Ni NWs, see Figure 7d. Therefore we neglect the magnetocrystalline anisotropy in the first approximation and the resulting magnetic field in the sample is expressed as

$$\mathbf{B}_{eff} = \mathbf{B} + \mathbf{B}_{demagn}, \tag{4}$$

where $\mathbf{B}_{demagn}$ is the induction of the demagnetized magnetic field mostly due to the shape anisotropy of the NWs. Under such assumptions the free energy in the external magnetic field B for the homogeneously magnetized system of Ni NW in PS will be

$$E = -\frac{\mathbf{M}\,\mathbf{B}}{\mu_0} - \frac{\mathbf{M}\mathbf{B}_{demagn}}{\mu_0} =$$

$$-\frac{Msin\theta Bsin\theta_B cos(\phi - \phi_B) + Mcos\theta Bcos\theta_B - MB_{demagn}sin^2\theta}{\mu_0} = \tag{5}$$

$$= -\frac{MB[sin\theta\ sin\theta_B\ cos(\phi - \phi_B) + cos\theta\ cos\theta_B] - MB_{demagn}sin^2\theta}{\mu_0}.$$

It was obtained that when the sample is rotated in an external magnetic field around an axis parallel to the normal of the PS template surface, the resonant

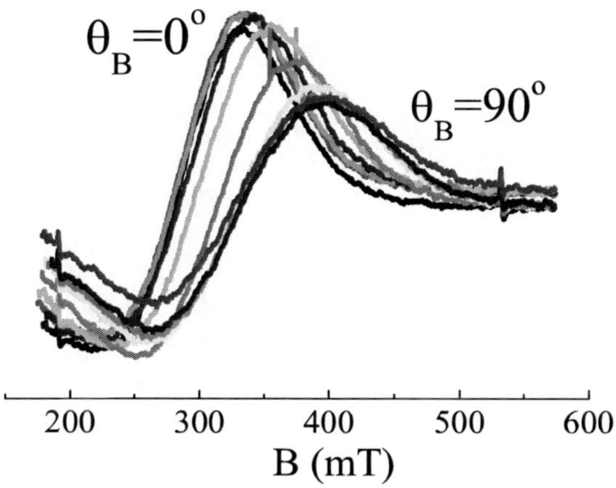

Figure 16. Measured FMR spectra of the array of Ni NWs in PS with $P = 0.7$ for different values of θ_B. From left to right, $\theta_B = 0^o, 10^o, 15^o, 30^o, 45^o, 60^o, 75^o, 80^o, 85^o, 90^o$.

magnetic field induction B_r remains the same. This fact testifies the IP isotropy of the magnetic properties of the samples, which is a consequence of the OOP orientation of Ni NWs grown in the PS templates. Therefore, if we choose the system of a spherical coordinates in such a way that in the $(x,\ y)$ plane the magnetic properties are isotropic (NWs are aligned along the z coordinate), then the angle $\phi = \phi_B$, and the expression (5) will reduce to

$$E = -\frac{MBcos(\theta_B - \theta) - MB_{demagn}sin^2\theta}{\mu_0} \tag{6}$$

The value of the angle θ will be determined by the condition of the minimum of the free energy

$$E_\theta = \frac{\partial E}{\partial \theta} = 0 \tag{7}$$

Combining Equations 6 and 7 we get

$$Bsin(\theta_B - \theta) + 2B_{demagn}sin\theta cos\theta =$$

$$= Bsin(\theta_B - \theta) + 2B_{demagn}sin(2\theta) = 0 \tag{8}$$

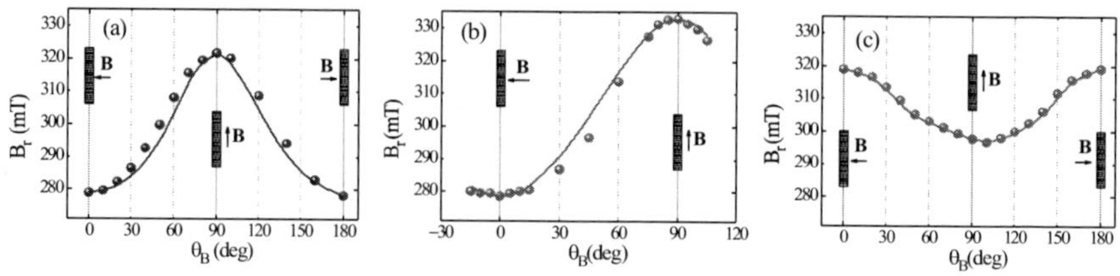

Figure 17. Dependence of the resonant magnetic inductance B_r on the angle θ_B for the arrays of Ni NWs in PS with (a) $P = 0.5$, (b) $P = 0.7$ and (c) $P = 0.8$.

Substituting Equation 6 for the free energy into Equation 1 we arrive to the condition of the resonance absorption of microwave field by system of Ni NWs in the external magnetic field

$$\left(\frac{\omega}{\gamma}\right)^2 = [B_r cos(\theta_B - \theta) + B_{demagn} cos(2\theta)][B_r cos(\theta_B - \theta) + B_{demagn} cos^2\theta], \tag{9}$$

where angle θ follows from the Equation 8. The negative value of B_{demagn} corresponds to the "easy plane" type of anisotropy, while the positive value of B_{demagn} is observed for the "easy axis".

Values of g-factor and demagnetized fields for the arrays of Ni NWs in the PS templates of different porosity can be assessed by solving the system of equations for the resonant field for limiting θ_B values, $\theta_B = 0^o$ and $\theta_B = 90^o$. On the base of (8) and (9) we get the following system of equations

$$\begin{cases} hf = g\mu_B(B_{OOP} + B_{demagn}), \\ (hf)^2 = g^2\mu_{\rm B}^2 B_{IP}(B_{IP} - B_{demagn}), \end{cases} \tag{10}$$

where B_{OOP} and B_{IP} are the resonant fields for OOP ($\theta = 0^o$) and IP ($\theta = 90^o$) configurations, respectively.

In Figure 17 the $B_r(\theta_B)$ dependencies for samples with different porosity ($P = 0.5$, 0.7 and 0.8) are shown by symbols. Lines represent the result of the best fit procedure according to the elaborated model, Equations 9 and 10. From that the values of g-factor and B_{demagn} were obtained. These results are summarized in Table 1.

The demagnetizing magnetic field can be decomposed into two components: the demagnetizing field due to the elongated shape of the NWs and the field due

Table 1. Values of g-factor, volume filling factor x_V, resulting B_{demagn} and type of the magnetic anisotropy for the arrays of Ni NWs in PS with different porosity P

Porosity	g-factor	x_V	B_{demagn}, mT	type of the magnetic anisotropy
0.5	2.14	0.29	+31.8	easy axis
0.7	2.12	0.29	+35.9	easy axis
0.8	2.11	0.34	-3.0	easy plane

to the flat shape of the PS template. Since the ratio of the diameter of individual NW to its length is much smaller than one, the demagnetizing field of the NW will be directed perpendicular to its axis as in the case of an infinitely long cylinder [61]. Thus, the first term of the demagnetization field will condition the perpendicular magnetic anisotropy with respect to the plane of PS template filled with nickel (the easy magnetization axis is parallel to the normal to the plane of the sample). On the other hand, Ni NWs are fairly close to each other, which leads to the presence of dipole-dipole interaction between them. Since the system of magnetic NWs in the PS template represents a plate with a thickness equal to the length of the NWs, which is much less than the in plane dimensions of the plate, the demagnetizing field directed perpendicular to the PS template surface will be present in samples as in the case of a thin plate (hard magnetization axis is parallel to the normal to the sample plane) [61]. As a result, two competing types of magnetic anisotropy will be present simultaneously in the samples: the easy axis, due to the extended structure of magnetic NWs, and the easy plane, which is a consequence of the location of these NWs in a flat template.

If we denote the demagnetization factors of individual NW in the plane and out of the plane directions as N_{IP} and N_{OOP}, respectively, ($2N_{IP} + N_{OOP} = 1$), in the approximation of the effective medium and assuming that all Ni NWs have the same sizes, the amplitude of the demagnetizing magnetic field B_{demagn} can be represented as follows

$$B_{demagn} = (1 - x_V)(N_{IP} - N_{OOP})\mu_0 M_{sat} - x_V \mu_o M_{sat}, \quad (11)$$

where x_V is the volume fiiling factor ($x_V = \frac{V_{NW}}{V_{sample}}$), where V_{NW} is the volume filled by NWs and V_{sample} is the overal volume of the porous template. Using

Equation 11 and the obtained B_{demagn} values it is possible to obtain the important parameter, x_V. Because the aspect ratio of the NWs is much greater than 1, it is reasonable to assume for a separate nanowire $N_{IP} = 0.5$ and $N_{OOP} = 0$ as for infinitely long cylinder. Substituting the value of the saturation magnetization of nickel with values previously obtained in section 4.1, $M_{sat} = 4 \times 10^5$ A/m, we obtain $x_V = 0.29$. Knowing the volume filling factor of nickel in the PS template and the porosity of the sample $P = 0.7$, one can estimate the upper limit of the degree of pore filling by metal, P_{full},

$$P_{full} \leq \frac{x_V}{P} \tag{12}$$

Note that the values of P_{full} and x_V were obtained under the assumption of the uniform filling of the pore length and homogeneous distribution of the nanowires in the sample. As a consequence, these values could be considered as the upper limit of the quantities. In addition, from Equation 11 it follows that the crossover of the magnetic anisotropy from the "easy axis" to the "easy plane" will occur under the following condition:

$$\frac{x_V}{(1 - x_V)(N_{IP} - N_{OOP})} > 1. \tag{13}$$

In particular, for the long cylinders ($N_{IP} = 0.5$, $N_{OOP} = 0$) the crossover in type of the anisotropy occurs at $x_V > 1/3$ [62, 63, 64]. In our case we observed the "easy plane" anisotropy only for the sample with $P = 0.8$, for which $x_V > 0.33$, see Table 1. This is in very good agreement with the theoretical prediction [63].

4.3. Isothermal Magnetic Hysteresis Loops

The isothermal $M(H)$ dependences, zero field cooled (ZFC) and field cooled (FC) magnetizations as functions of T were measured using a MPMS XL SQUID (Quantum Design). The magnetic field was varied in the range from -2 to $+2$ T, within the temperature range 2–350 K. The magnetic moment was measured with the sensitivity of 10^{-8} emu. Magnetic field was applied either parallel (OOP configuration) or perpendicular (IP configuration) to the Ni NW axis. All the results presented in this Section were obtained on the PS templates with $P = 0.7$ and $t_{Ni} = 60$ min.

The magnetization curves, recorded for OOP and IP configurations at two temperatures ($T = 5$ K and 300 K), are presented in Figure 18. The evolution

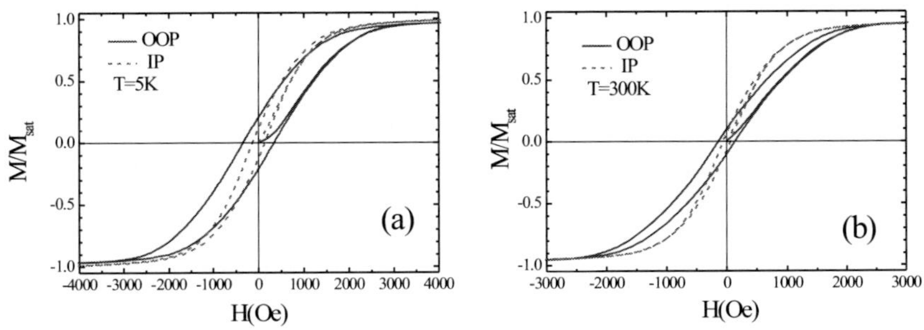

Figure 18. Field dependences of the normalized magnetization M/M_{sat} for the array of Ni NWs in PS. (a) $T = 5$ K, (b) $T = 300$ K.

of the magnetic hysteresis loops shows a narrowing tendency towards higher temperatures.

Three main magnetostatic parameters can be extracted from the isothermal hysteresis magnetization curves, the saturation magnetization M_{sat}, the coercivity H_C and the remanent magnetization M_{rem}, as well as their temperature dependences for the both configurations.

In Figure 19 we show the temperature dependence of the squareness M_{rem}/M_{sat} of the magnetic hysteresis loops and coercivity H_C for both orientations of the magnetic field. Circles are for the OOP and squares refer to the IP configurations. Closed and open symbols refer to squareness and coercivity, respectively. It can be seen that the "easy axis" anisotropy dominates for the OOP configuration in the whole temperature range, including the room temperature, which corresponds to the results of the FMR study, see Figure 17b. On the other hand, the saturation field H_{sat} tends to be less for the IP configuration, see Figure 18. It means that for the IP orientation of the magnetic field the sample reaches the saturation magnetization at lower magnetic field values. For Ni the critical diameter below which the nanoparticle becomes single domain, is equal to $D_c = 55$ nm [65]. This value is very close to the diameter of NWs of this research ($\approx 60 - 80$ nm). Therefore, it is reasonable to assume that Ni NWs in the PS templates consist of polycrystalline material with single domain grains. It means that, in the IP direction NW consists almost of one domain, while for the OOP direction the number of domains which should be aligned is much higher, around 160.

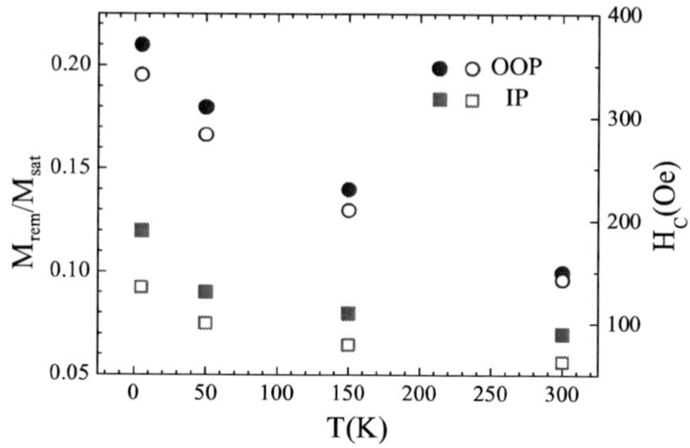

Figure 19. Temperature dependence of the squareness M_{rem}/M_{sat} of the magnetic hysteresis loops and coercivity H_C for the array of Ni NWs in PS. Closed (open) symbols refer to squareness (coercivity).

The ZFC-FC curves reveal the typical features of an ensemble of ferromagnetic particles interacting with different strength with respect to the orientation of the magnetic field [56]. In Figure 20 the ZFC-FC magnetization curves are shown for the both orientations. Data were obtained with an external magnetic field of 100 Oe. It is clearly seen that curves only coincide at the highest temperature measured (350 K) confirming that the blocking temperature (T_B) is higher than this T. Nevertheless, it is possible to extract useful qualitative information from the analysis of these data. The general increase with temperature of the ZFC branch indicates a dipolar coupling as between grains in NWs as well between NWs. A near flat FC curve for OOP configuration at $T>100$ K indicates the expected strong interaction of domains along the NW axis. For IP configuration the FC magnetization increases for $T>100$ K and at $T>270$ K it even exceeds the value of magnetization at liquid helium temperature. This could be the signature of strong dipole-dipole interaction of NWs within plane. Moreover, the visible difference in the ZFC-FC measurements is defined by the magnetic anisotropy of the sample. Because the material has a magnetic anisotropy, the values of the magnetization obtained in the field 100 Oe (which is much less than the saturation field H_{sat}) in the IP and OOP geometries vary. This difference can be seen on virgin curves of the magnetization for the IP and OOP configuration (see Figure 18).

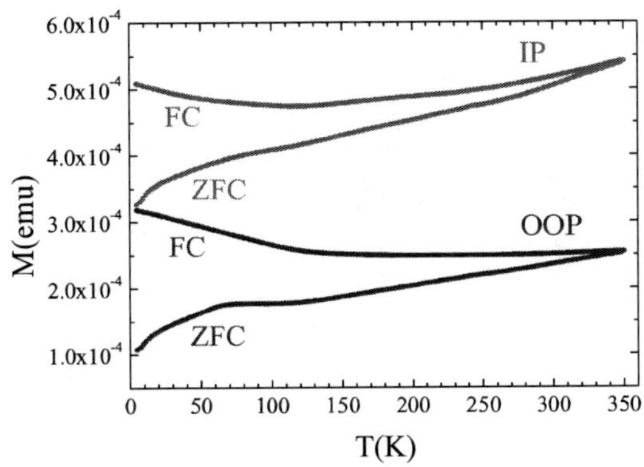

Figure 20. ZFC and FC magnetizations as functions of temperature for the array of Ni NWs in PS.

Additional important information about magnetostatic interaction between NWs was obtained by applying magnetic force microscopy (MFM) in combination with atomic force microscopy (AFM). The XE-120 scanning probe microscope (SPM) from Park System Corporation was used for both AFM and MFM imaging. Non-contact topography scans and MFM measurements were carried out with "MFMR" cantilevers (NanoWorld AG) covered with 40 nm thick hard magnetic cobalt alloy and having permanent tip magnetization direction along the tip axis. The radius of tip curvature was 50 nm. The SPM system was equipped with 100×100 μm^2 XY-scanner and 25 μm Z-scanner. All SPM scans were performed at ambient conditions. Before MFM scans samples were demagnetized by applying a diminishing alternating magnetic field along the "easy axis" of magnetization. Shifts in the phase of cantilever oscillation were measured in phase detection mode to examine the magnetic activity on the samples surface. The surface morphology and magnetic features of the PS template filled with Ni are shown in Figure 21. Based on the AFM results the observed structures are typically $60 - 80$ nm large, which agrees well with the average pore diameters obtained by SEM. The MFM image were achieved after topography measurement followed by sample surface scanning at a constant 300 nm height. The comparison of the AFM/MFM measurements clearly shows that the magnetic moments have OOP component and the mean size of magnetic domain (approximately 1 μm) is much larger than the diameter of the pores. It

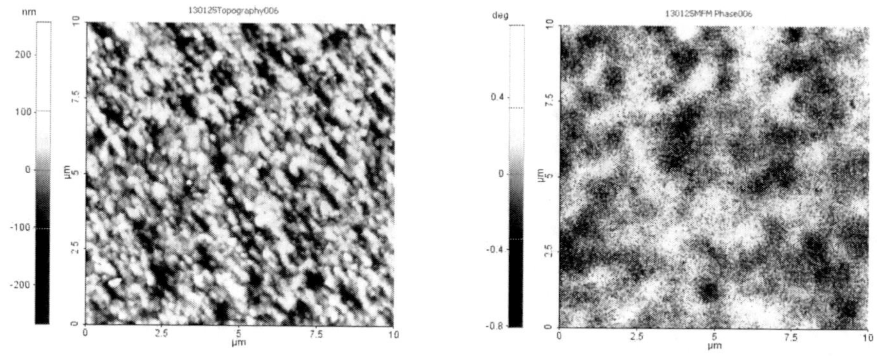

Figure 21. AFM (left side) and corresponding MFM (right side) images of the PS template filled with Ni. The phase contrast of magnetic signal was measured at 300 nm tip-sample distance.

indicates that magnetic domain consist of several NWs with the same magnetization direction. This result is in good agreement with the previously described behaviour where the "easy axis" orientation and strong coupling between NWs were observed.

Now we discuss the coercivity values and coercivity anisotropy of Ni NWs. Coercivity is one of the most important properties of magnetic materials determining different applications like magnetic recording and spin electronics. The measured at room temperature H_C values are 143 Oe for OOP and 63 Oe for IP configurations, much higher the coercivity of bulk Ni (around 0.7 Oe for Ni) [66].

It is well known that both number and size of domains play a major role in magnetic properties [61]. The values of H_C, for nanoscale samples usually are much larger than that of bulk, and it decreases with increasing size due to the transition from single domain to multidomain regime. As the size of nanostructure increases, coercivity decreases due to the increase in the number of domain walls. However, nanoscale samples below D_c, having a single magnetic domain, are considerably affected by the shape and surface anisotropy energies, besides the thermal fluctuations. For this range of sizes H_C increases by a factor of two as the diameter of Ni NWs increases from 10 to 20 nm indicating that the magnetic properties are rather sensitive at nanoscale [67].

For single domain nanoparticles and for intermediate temperatures H_C is expected to follow the law [56]

$$H_C(T) = H_C(0)\left[1 - \left(\frac{T}{T_B}\right)^{\zeta}\right], \tag{14}$$

where T_B is the blocking temperature of the largest particles in the system and the exponent ζ depends on the alignment of the particles [56]. For an assembly of aligned particles $\zeta = 0.5$ [68], while for randomly oriented particles $\zeta = 0.77$ [69]. In Figure 22 we show the dependence of the coercivity H_C on $T^{0.5}$ values of the temperature for the both configurations. It follows from Figure 22 that this dependence is almost linear. Analysis of the experimental data with the exponent $\zeta = 0.77$ gives a worse agreement, see inset to Figure 22.

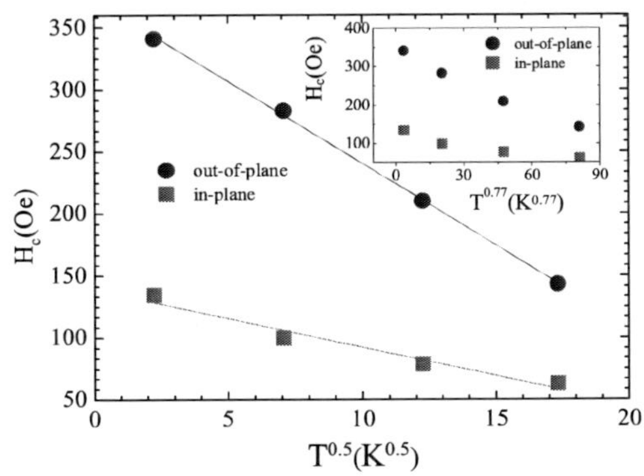

Figure 22. The coercivity H_C versus $T^{0.5}$ for the array of Ni NWs in PS. Circles (squares) refer to the OOP (IP) configuration. Solid lines represent the best fit to the experimental data. Inset: The coercivity H_C versus $T^{0.77}$ for the array of Ni NWs in PS. Circles (squares) refer to the OOP (IP) configuration.

On the basis of the obtained experimental data it is possible to clarify the mechanism of magnetic interaction within polycrystalline Ni NWs. For that we consider here the random anisotropy model (RAM) which was first proposed by Harris et al. [70] and compare its theoretical predictions with our results obtained from the magnetization measurements. The RAM has been successfully applied in the past to explain the properties of both amorphous and nanocrystalline ferromagnets [71, 72, 73, 74]. The key to the RAM is the assumption of the presence of disorder in the direction of the local magnetic anisotropy and the possibility of describing the magnetic structure of a set of weakly coupled magnetic blocks of size of ferromagnetic correlation length R_F [75]. The law of the approach to saturation (LAS) reflects the mechanism of this disorder. The general simple expression for the LAS, which consider the dimensionality d of the nanocrystalline ferromagnet, was derived in the work [73]

$$\frac{\delta M(M)}{M_{sat}} = \frac{M_{sat} - M(H)}{M_{sat}} = \left(\frac{1}{\sqrt{15}} \times \frac{< H_a >}{H}\right)^{-\frac{4-d}{2}} \sim H^{-\frac{4-d}{2}}, \quad (15)$$

where $< H_a >$ is the average anisotropy field and the quantity d determines the dimensionality of the system. In particular, for single domain non interacting nanoparticles $d = 0$, while for strong interacting nanoparticles $d = 3$, which leads to the dependence $\delta M(H)/M_{sat} \sim H^{-1/2}$ [73]. The behavior (15) is valid for uniaxial amorphous or nanocrystalline magnetic materials.

The magnetization curves for all temperatures are found to fit well Equation 15 as shown in Figure 23. Here for simplicity we present the result only for 2 temperatures, $T = 5$ K and $T = 300$ K. Data are for OOP configuration.

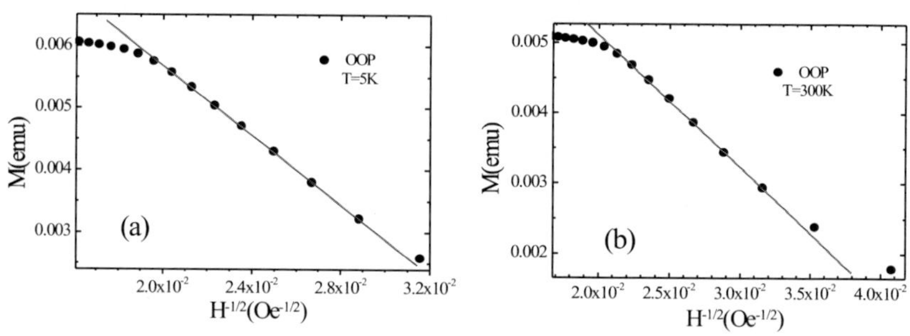

Figure 23. Analysis of LAS for the array of Ni NWs in PS. OOP configuration. Symbols are for the experimental data. Solid lines are for the best fit results according to the Equation (15). (a) $T = 5$ K, (b) $T = 300$ K.

It is seen that the experimental data are well described by the law $\delta M/M_{sat} \sim H^{-1/2}$, which means $d = 3$. Such LAS is usually reported when the exchange coupling between nanoparticles dominates. From the slope of the solid lines in the Figure 23 it is possible to extract the average anisotropy value, $< H_a >$, at different temperatures. In Figure 24 we plot the temperature dependencies of $< H_a >$ for the both configurations. It follows from this result, that the absolute values of the anisotropy field are less for the IP configuration in the whole investigated temperature range. This coincides with the previously obtained results regarding the "easy axis" orientation in Ni NWs in PS with $P = 0.7$.

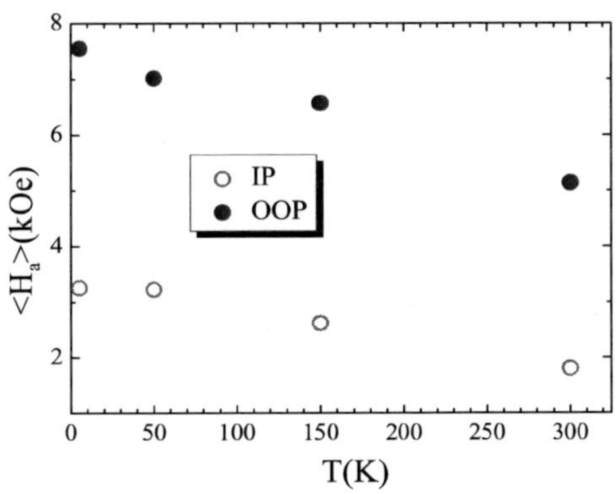

Figure 24. The average anisotropy magnetic field $< H_a >$ versus temperature for OOP (closed symbols) and IP (open symbols) configurations.

5. Conclusion

In this chapter we have presented some of the characteristics of Ni ferromagnetic NWs grown by electrodeposition into the PS templates of very good uniformity. We have developed the model of growth of continuous Ni NWs, taking into account the kinetics and importance of the interaction between solution and pore walls. The role of amorphous a-Si in nickel silicides formation during the Ni NWs growth was explicitly established. The presence of Ni_2Si can be controlled by the heat treatment in vacuum thus allowing the tuning of the specific magnetization and Curie temperature of Ni NWs.

The uniaxial anisotropy was established from the FMR study. By analyzing the angular dependence of the resonant magnetic field at $f = 9.32$ GHz for arrays of Ni NWs in PS with different porosity we observed the crossover from the "easy axis" magnetization for $P \leq 0.7$ to the "easy plane" at $P>0.7$. Such important parameters as g-factor and the demagnetizing field were also evaluated from these measurements, as well as the volume filling factor.

From the magnetostatic measurements the evidence of the "easy axis" oriented along the NW axis was confirmed. At that the analysis of LAS revealed that, the anisotropy field is greater for the OOP configuration with the respect to the IP, which confirms the orientation of the "easy axis" along the NW.

Acknowledgments

The authors would like to thank V. Tsybulsky for SEM analysis, A.Mazanik, I. Komissarov and O. Korolik for Raman spectroscopy measurements, S. Redko for technical assistance during the sample fabrication. Authors from Belarus acknowledge the financial support from the Belarusian Government Research Program "Functional Engineering Materials, Nanomaterials" under grant no. 2.4.12. V. Bondarenko acknowledges professor R.Herino for paying his attention to the problem of the electrochemical deposition of metals in PS in 1986.

References

[1] Liu K, Nagodawithana K, Searson P, Chien C, *Phys Rev B* 51, 7381-7385 (1995).

[2] Tsoi M, Sun J, Rooks M, Koch R, Parkin S, *Phys Rev B* 69, 100406(R)(1-4) (2004).

[3] McGary P D, Tan L, Zou J, Stadler B J H, Downey P R, Flatau A B, *J Appl Phys* 99, 08B310(1-6) (2006).

[4] Ramos C A, Brigneti E V, De Biasi E, Vázquez M, in *Nanowires - Fundamental Research* ed. Hashim A (InTech, 2011).

[5] Cheng-Zhang L, Lodder J C, *J Magn Magn Mater* 88, 236-246 (1990).

[6] Spiegel J, Huynen I, *Solid State Pnenomena* 152-153, 389-393 (2009).

[7] Mehta K K, Wu T-H, Chiou E P Y, *Appl Phys Lett* 93, 254102(1-3) (2008).

[8] Gao N, Wang H J, Yang E H *Nanotechnology* 21, 105107(1-8) (2010).

[9] Choi D S, Park J, Kim S, Gracias D H, Cho M K, Kim Y K, Fung A, Lee S E, Chen Y, Khanal S, Baral S, Kum J-H, *J Nanosci Nanotechnol* 8, 2323-2327 (2008).

[10] Chou S Y, Krauss P R, Kong L, *J Appl Phys* 79, 6101-6106 (1996).

[11] Grobert N, Hsu W K, Zhu Y Q, Hare J P, Kroto H W, Walton D R M, Terrones M, Terrones H, Redlich Ph, Rühle M, Escudero R, Morales F, *Appl Phys Lett* 75, 3363-3365 (1999).

[12] Dinner R B, Robinson A P, Wimbush S C, MacManus-Driscoll J L, Blamire M G, *Superc Sci Tecnhol* 24, 055017(1-8) (2011).

[13] Harraz F A, Sakka T, Ogata Y H, *Phys Status Solidi A* 197, 51-56 (2003).

[14] Kawai S, Ueda R, *J Electrochem Soc* 122, 32-36 (1975).

[15] Ye Z, Liu H, Luo Z, Lee H, Wu W, Naugle D, Lyuksyutov I, *Nanotechnology* 20, 045704(1-8) (2009).

[16] Whitney T M, Searson C, Jiang J S, Chien C L, *Science* 261, 1316-1319 (1993).

[17] Piraux L, Dubois S, Ferain E, Legras R, Ounadjela K, George J M, Maurice J L, Fert A, *J Magn Magn Mater* 165, 352-355 (1997).

[18] Cui W, Lu X, Su B, Lu Q, Wei Y, *Appl Phys Lett* 95, 153102(1-3) (2009).

[19] Yeh W J, Yang K, *J Phys Conf Ser* 61, 1299-1302 (2007).

[20] Gusev S A, Korotkova N A, Rozenstein D B, Fraerman A A, *J Appl Phys* 76, 6671-6672 (1994).

[21] Dolgiy A L, Redko S V, Komissarov I, Bondarenko V P, Yanushkevich K I, Prischepa S L, *Thin Solid Films* 543 133-137 (2013).

[22] Granitzer P, Rumpf K, *Materials* 4, 908-928 (2011).

[23] Uhlir A, *Bell System Technical Journal* 35, 333-347 (1956).

[24] Tsao S, Blewer R, Tsao J, *Appl Phys Lett* 49, 403-405 (1986).

[25] Herino R, Jan P, Bomchil G, *J Electrochem Soc* 132, 2513-2514 (1985).

[26] Borisenko V E, Bondarenko V, Raiko V, *Doklady Academy Sciences Belarus* 33, 538-540 (1989).

[27] Hamadache F, Renaux C, Duvail J, Bertrand P, *Physics Status Solidi (a)* 197, 168-174 (2003).

[28] Kimura T, Yokoi A, Horiguchi H, Saito R, Ikoma T, Sato A, *Appl Phys Lett* 65, 983-985 (1994).

[29] Bondarenko V, Dolgyi L, Dorofeev A, Kazuchits N, Leshok A, Troyanova G, Vorozov N, Maiello G, Masini G, La Monica S, Ferrari A, *Thin Solid Films* 297, 48-53 (1997).

[30] Herino R, in: *Properties of Porous Silicon* Ed. L. Canham. Ch. 2.2 (INSPEC, London) (1997).

[31] Rusetskii M S, Kazyuchits N M, Bayev V G, Dolgii A L, Bondarenko V P, *Pisma Zh. Techn. Fiz.* 37, 1-6 (2011). [*Tech. Phys. Letters* 37, 391-396 (2011)].

[32] Aravamudhan S, Luongo K, Poddar P, Srikanth H, Bhansali S, *Appl. Phys. A* 87, 773-780 (2007).

[33] Granitzer P, Rumpf K, *Materials* 3, 943-998 (2010).

[34] Michelakaki E, Valalaki K, Nassiopoulou A G, *J Nanopart Res* 15, 1499(1-9) (2013).

[35] Granitzer P, Rumpf K, Ohta T, Koshida N, Reissner M, Poelt P, *Appl Phys Lett* 101, 033110(4) (2012).

[36] Dolgyi A, Bandarenka H, Prischepa S, Yanushkevich K, Nenzi P, Balucani M, Bondarenko V, *Electrochem Soc Trans* 41, 111-118 (2012).

[37] Dolgiy A, Redko S V, Bandarenka H, Prischepa S L, Yanushkevich K, Nenzi P, Balucani M, Bondarenko V P, *J Electrochem Soc* 159, D623-D627 (2012).

[38] Panarin A Yu, Terekhov S N, Kholostov K I, Bondarenko V P, *Appl Surf Sci* 256, 6969-6976 (2010).

[39] International Centre for Diffraction DATA, *PCPDFWIN, JCPDS*, 2, 89-838; 85-5648.

[40] Pan H, Sun H, Poh C, Feng Y, Lin J, *Nanotechnology* 16, 1559-1564 (2005).

[41] Tu K N, Ottaviani G, Gösele U, Föll H, *J Appl Phys* 54, 758-763 (1983).

[42] Borisenko V E, *Semiconducting Silicides* (Springer) 348 p, (2000).

[43] Tu K N, *Appl Phys Lett* 27, 221-224 (1975).

[44] van Loeneu E J, Frenken J W M, van der Veen J F, *Appl Phys Lett* 45, 41-43 (1984).

[45] Zunger A, *Phys Rev B* 24, 4372-4391 (1981).

[46] Desimioni J, Traverse A, *Phys Rev B* 48, 13266-13272 (1993).

[47] Boussaa N, Guittoum A, Tobbeche S, *Vacuum* 77, 125-130 (2005).

[48] Canham L T, *Appl Phys Lett* 57, 1046-1048 (1990).

[49] Voutsas A T, Hatails M K, Boyce J, Chiang A, *J Appl Phys* 78, 6999-7006 (1995).

[50] Kim H-S, Yong Y-J, Lee J-Y, Lee K-Y, *J Mater Sci Lett* 16, 560-562 (1997).

[51] Lien C-D, Nicolet M-A, *Thin Solid Films* 143, 63-72 (1986).

[52] Mangelinc D, Hoummada K, *Appl Phys Lett* 92, 254101(1-3) (2008).

[53] Ravinskii A F, Makoed I I, Kokoshkevich K, Yanushkevich K I, Galyas A I, Triguk V V, *Inorg Mater* 43, 860-865 (2007).

[54] Pauthenet R, in *High Field Magnetism*, ed. Dafe M, (North-Holland Publ Comp, Amsterdam) 77 p (1983).

[55] Ashkroft N W, Mermin N D, *Solid State Physics* (North Holland, Amsterdam), vol. 2 (1992).

[56] Dormann J L, Fiorani D, Tronc E, *Adv Chem Phys* 98, 283-494 (1997).

[57] Weil J A, Bolton J R, Wertz J E, *Electron paramagnetic resonance Elementary theory and practical applications*, (New York, Wiley-Interscience) 592p (1994).

[58] Poklonski N A, Gusakov G A, Bayev V G, Lapchuk N M, *Semiconductors* 43, 568-576 (2009).

[59] Smit J, Beljers H G, *Philips Research Reports*, 10, 113-130 (1955).

[60] Landau L D, Lifshits E M, *Electrodynamics of Continuous Media* (Pergamon Press, Oxford, England) 2nd edition (1984).

[61] Cullity B D, Graham C D, *Introduction to Magnetic Materials* (John Wiley & Sons : IEEE Press, Hoboken, New Jersey) 544 p. (2009).

[62] Encinas-Oropesa A, Demand M, Piraux L, Ebles U, Huynen I, *J Appl Phys* 89, 6704-6706.

[63] Carignan L-P, Lacroix C, Ouimet A, Ciureanu M, Yelon A, Ménard D, *J Appl Phys* 102, 023905(1-10) (2007).

[64] Kartopu G, Yalcin O, Choy K-L, Topkaya R, Kazan S, Aktas B, *J Appl Phys* 109, 033909(1-8) (2011).

[65] Chernavskij P A, *Russian Chemical Journal* XLVI, 19-30 (2002).

[66] Chikazumi S, *Physics of Ferromagnetism, Magnetic Properties of Materials* (Syokabo, Tokyo) (1980).

[67] Zheng M, Menon L, Zeng H, Liu Y, Bandyopadhyay S, Kirby R D, Sellmyer D J, *Phys Rev B* 62, 12282-12286 (2000).

[68] Bean C P, Livingston J D, *J Appl Phys* 30, S120-S129 (1959).

[69] Pfeiffer H, Schüppel W, *Phys Status Solidi A* 119, 259-69 (1990).

[70] Harris R, Pilschke M, Zukermann M J, *Phys Rev Lett* 31, 160-162 (1973).

[71] Tejada J, Martinez B, Labarta A, Grössinger R, Sassik H, Vázguez M, Hernando A, *Phys Rev B* 42, 898-905 (1990).

[72] Löffler J F, Meier J P, Doudin B, Ansermet J-P, Wagner W, *Phys Rev B* 57, 2915-2924 (1998).

[73] Iskhakov R S, Komogortsev S V, Balaev A D, Chekanova L A, *Pisma Zh Eksp Teor Fiz* 72, 440-444 (2000). [*JETP Letters* 72, 304-307] (2000).

[74] Yamkane Z, Lassri H, Menai A, Khazzan S, Mliki N, Bessais L, *J Alloys Comp* 584, 352-355 (2014).

[75] Chudnovsky E M, *J Appl Phys* 64, 5770-5775 (1988).

In: Nanowires
Editor: Luke J. Wilson, pp. 129-152
ISBN: 978-1-63117-855-9

Chapter 4

PHONON SCATTERING AND ELASTIC ENERGY PROPAGATION IN NANOWIRES

E. I. Salamatov*
Physical Technical Institute UrB RAS, Izhevsk, Russia

Abstract

The finite-difference time-domain (FDTD) method is used to study the long-wave phonon scattering at geometric imperfections of two-dimensional nanowires. The transmission/reflection spectra of the phonons of different polarizations (bending and compression phonons) are calculated for two kinds of defects: retaining and not retaining the waveguide symmetry. These calculations are relevant to the description of ballistic phonon thermal transport in mesoscopic systems at low temperatures. The elastic energy propagation in a nanowaveguide between two massive blocks is studied by FDTD method. The energy field realized in the system in the process of formation and propagation of the energy flow is calculated. It is shown that in mesoscopic systems the field configuration is determined by the sample geometry and the boundary conditions. Based on the calculations and analysis performed, the conclusion is drawn that in the steady-state regime of energy propagation between two massive blocks through a small-section waveguide with a free surface, energy propagates in a narrow surface layer.

Keywords: FDTD method, low-dimensional systems, nanowires, mesoscopic systems, phonon scattering, elastic energy propagation

*E-mail address: esalama2i@gmail.com

1. Introduction

The present-day interest in heat transport by phonons in mesoscopic systems is determined by general tendency of electro-acoustic devices to shrink in size because of the appearance of advanced technologies that permit to control, with almost atomic precision, the geometry of a sample. Direct measurements of the thermal conduction of nanometer waveguides in mesosystems comparable in size to the dominant phonon wavelength at a temperature of 1K ($\sim 100nm$) were first carried out in [1]. In early theoretical works initiated by those experiments [2-4] particular attention was paid to quantum effects of thermal conduction in the process of ballistic phonon propagation in nanowaveguides. Presently, special attention is given to the possibility of manipulating the heat flow in technological devices. Along with traditional techniques of controlling the thermal energy flow based on introducing impurities, defects, and alloy atoms in the crystal lattice, many alternative methods have been developed and several nanostructures and complex compounds have been used to control heat transport by means of guest atoms and phonon interface scattering [5-7]. In mesoscopic systems the heat flow is strongly affected by the sample geometry. In particular, it was shown that the contact shape acts as an acoustic impedance adaptor playing a crucial role in the ballistic phonon propagation and thermal conduction [8]. We believe that the energy field configuration which in mesoscopic systems is greatly determined by surface effects due to both the sample geometry and boundary conditions, should influence the phonon (elastic wave) propagation.

The theoretical and numerical approaches are widely used to interpret the dynamical and transport characteristics of nanowires. The most suitable for this purpose theoretical models are the Green's function [9-12] and the transfer matrix [13-16] methods. The plane wave expansion (PWE) method is commonly used to study the phonon spectra of infinite periodic structures such as phonon lattices [17]. However, the PWE method cannot be applied directly in the case of samples with a finite cross section, since it assumes either an infinite structure or periodic boundary conditions, and in this case the finite element method [18] is appropriate. In this work we use the finite-difference time-domain (FDTD) method [19] which allows us to investigate finite systems, simulating the experiments in the same way they are carried out and making comparison. At each instant of time it provides the displacement field at every point making it possible to study in real time the response of the defect to the perturbance. The

FDTD method, developed in the 1970s for purposes of geophysics to describe the seismic wave propagation [20], is widely used today for simulation of both electromagnetic [21] and elastic [22-25] waves propagation in inhomogeneous solids. The advantage of this method over other techniques is that it can be used in investigating the wave propagation in specimens of inhomogeneous composition, complex geometry, and with different boundary and initial conditions. Within the FDTD method one can study a unit event of scattering of an elastic wave (phonon) at arbitrarily complex defects, and investigate the distribution and flow of elastic energy.

In the present work we study the scattering of long-wave acoustic phonons of different polarization (bending and compression waves) in nanowires at bifurcation points of the main nanowaveguide. Then, the kinetic energy of elastic vibrations (an analog of temperature)is considered when studying the evolution of the energy field configuration in the process of attainment of steady-state flow in a waveguide between two massive blocks. The results obtained may give a clearer insight into the process of elastic energy propagation in such mesoscopic finite-size systems. In addition, at low temperatures ($T < 1K$) the heat transfer in such systems is commonly believed to occur via ballistic phonons, and in many theoretical descriptions of the phonon (plane waves) scattering at the abrupt junction between the nanowaveguide and the blocks the elasticity theory is used [2,3,8,26,27], which gives promise that our results will be useful in interpreting the thermal conduction mechanism.

2. FDTD Method

2.1. Finite Systems

Consider a two-dimensional homogeneous nanowire, the length of which in the x-direction, L, far exceeds the cross-section b ($L \gg b$). The motion of lattice displacement $u(\mathbf{r}, \mathbf{t})$ in an elastic continuum is described by the equations

$$\rho(\mathbf{r})\partial^2 u_i(\mathbf{r}, t)/\partial t^2 = \partial_j \sigma_{ij}(\mathbf{r}, t), \tag{1}$$

$$\sigma_{ij}(\mathbf{r}, t) = c_{ijmn}(\mathbf{r})\partial_n u_m(\mathbf{r}, t), \tag{2}$$

where $\rho(\mathbf{r})$ and $c_{ijmn}(\mathbf{r})$ are, respectively, the mass density and the elastic stiffness tensor which in this work are considered to be constant, $\mathbf{r} = (\mathbf{x}, \mathbf{y})$, $\sigma_{ij}(\mathbf{r})$ is the stress tensor.

To solve these equations it is necessary to set the boundary and initial conditions. In the work two types of boundary conditions are used (except for the case considered below) :

1. The boundary condition at the calculated-cell boundaries ($x = 0$, $x = L$) in the x-direction is the Mur absorption condition [28]:

$$u_i(1, m, t+\Delta t) = u_i(2, m, t) + \frac{v_0\Delta t - \Delta x}{v_0\Delta t + \Delta x} \times (u_i(2, m, t+\Delta) - u_i(1, m, t)), \tag{3}$$

written for the left boundary of the cell. Here $u_i(l, m)$ ($i = x, y$) is the displacement of the unit element l, m of area $\Delta x\Delta y$, Δt is the time step, c_0 is the sound velocity.

2. The boundaries normal to the y-axis are assumed to be free. There are some ways of formalizing this condition for FDTD calculations [29]. In our work the vacuum formulation, convenient for objects of complex geometry, is used [30]. It consists in introducing an additional plane in vacuum, on which the necessary elements of the stress tensor are assigned such values that the linear approximation yields zero values of these components on the surface.

The initial conditions are chosen according to the task.

Once appropriate initial and boundary conditions are specified, Eqs. (1) and (2) can be solved numerically by discretizing both the time (Δt) and space ($\Delta x, \Delta y$) domains. In this work the elasticity equations are solved by the FDTD method realized on a staggered grid $N\Delta x * M\Delta y$. The spatial derivatives are calculated using the central-difference approximation and 2nd-order approximation to the second derivative with time. For the convergence of the procedure the condition $\Delta t < \Delta x/(2c_l)$ must be satisfied.

2.2. Infinite Periodic Systems

To analyze the results of the numerical simulation it is very useful to know the dispersion of phonons in an ideal infinite system. In this case, owing to the periodicity in the x direction, the lattice displacement and the stress tensor take the form satisfying the Bloch theorem

$$u_i(\mathbf{r}, t) = \exp(ik_x x)U(\mathbf{r}, t) \tag{4}$$

$$\sigma_{ij}(\mathbf{r}, t) = \exp(ik_x x)S_{ij}(\mathbf{r}, t), \tag{5}$$

where k_x is a Bloch wave vector and $U_i(x, t)$ and $S_{ij}(x, t)$ are periodic functions satisfying $U_i(x + a, t) = U_i(x, t)$ and $S_{ij}(x + a, t) = S_{ij}(x, t)$, with a lattice translation vector a. In numerical calculations of the elasticity theory problem the a value in the ideal system is arbitrary and usually equal to several Δx. Thus, Eqs. (1) and (2) are rewritten as

$$\rho(\mathbf{r})\partial^2 U_i(\mathbf{r},t)/\partial t^2 = ik_x S_{xj}(\mathbf{r},t), \tag{6}$$

$$S_{ij}(\mathbf{r},t) = c_{ijmx}(\mathbf{r}) ik_x U_m(\mathbf{r},t). \tag{7}$$

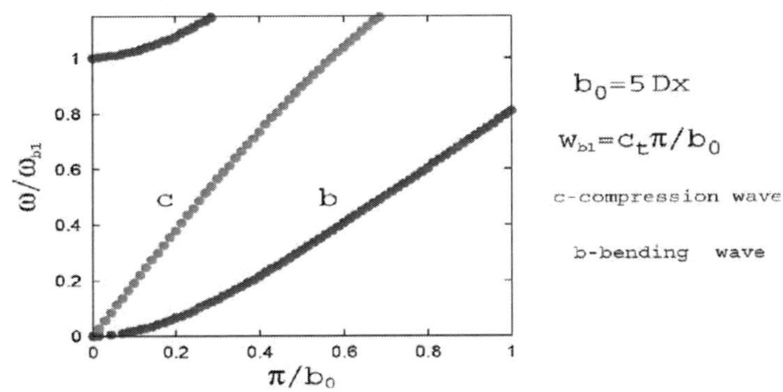

Figure 1. The phonon dispersion law for a two-dimensional ZrO_2 waveguide.

Now we try to solve these equations with respect to the reduced fields $\mathbf{U}$ $(\mathbf{r}, t)$ and $S_{ij}(x, t)$ within a unit cell of the structure. First we specify a wave vector k_x in the first Brillouin zone. Once appropriate initial conditions are specified, Eqs. (6) and (7) can be solved for each normal mode. The explicit expressions for the discretized versions of these equations are given , for example, in [31]. More explicitly, when the displacement fields are specified at an instant $t = 0$, their spatial derivatives are determined using simple finite-difference formulas. Equations (6) and (7) then give us the time derivative of the displacement field $\mathbf{U}$, which allows us to update $\mathbf{U}$ for small but positive t. In this way, the displacement fields $\mathbf{U}$ $(\mathbf{r}, t_i)$ at discretized points on the time axis $t_i (i = 1, 2, ...)$ are determined for many 2D grid points sampled. For a sufficiently large number of these $\mathbf{U}$ data on the time axis, the displacement fields are Fourier-transformed into the frequency space. The positions of the existing peaks in the frequency spectra are then identified as the eigenfrequencies of the normal vibrational modes for a given wave vector k_x.

The dispersion law of the infinite waveguides for ZrO_2 is presented in Figure 1. Two acoustic branches correspond to the bending waves antisymmetrical relative to the wire axis (b), and symmetrical compression waves (c), respectively.

In the long-wave limit the dispersion law for the acoustic compression and bending waves is $\omega_c(k) = c_t k$, $\omega_b(k) = ((1+\sigma)/6)^{1/2} 2 c_t b_0 k^2$, and the boundary frequencies of the lowest optical branches of the corresponding waves are equal to $\omega_{c1} = (1/(1-\sigma))^{1/2} \pi c_t / b_0$ and $\omega_{b1} = \pi c_t / b_0$ [26], where c_t is the velocity of the transversal wave in the bulk sample, σ is the Poisson's ratio. When calculating, the values for the density and elastic constants of zirconia were taken to be $\rho = 6.5 g/cm^3$ and (in units $10^{10} dyn/cm^2$) $c_{11} = 417$, $c_{44} = 47$ [32]. The elastically isotropic approximation ($c_{12} = c_{11} - 2c_{44}$) was used.

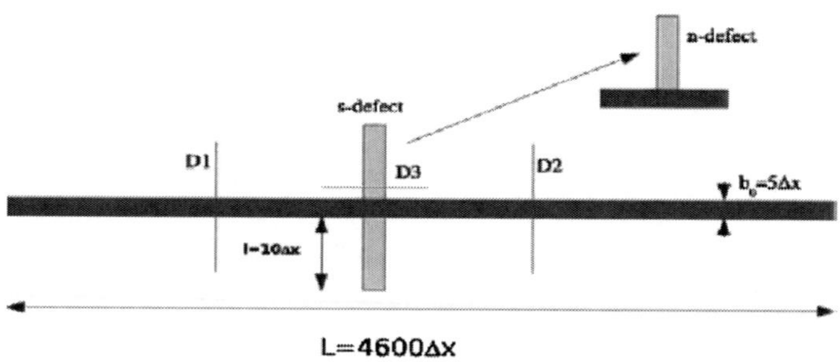

Figure 2. The calculated cell for the study of phonon scattering by defects.

All geometrical unit values in the work were normalized to the linear size of the unit cell ($\Delta x = \Delta y$) which determines the minimum length, $\lambda = 2\Delta x$, and correspondingly the maximum frequency, $\nu_{max} = c_l / 2\Delta x$, of the elastic wave, which can be described within the framework of the FDTD method. The time step was chosen to be $\Delta t \simeq 10^{-2} Tmin = 10^{-2}/\nu_{max}$. All the results will be presented in these units, so to give an idea of the order of magnitude, note that in the case of $\Delta x = 10nm$ (for such distances the elasticity theory is valid) $\nu_{max} = 4 * 10^{11} Hz$ $\Delta t = 0.6ps$ for zirconia. In modern devices the linear sizes of the system elements are $L = (1-5) * 10^{-6} m$, $b = (50 - 100) nm$, therefore in calculations the value of b varied from 4 to 15 Δx, and $L > 300\Delta x$. The frequencies in the work are expressed in units ω_{b1}.

3. Phonon Scattering at Geometric Imperfections in Nanowires

3.1. Calculation

The transmission/reflection spectra of the phonons of different polarization (bending and compression phonons) are calculated for two kinds of bifurcation points: symmetrical (s) and nonsymmetrical (n) defects (Figure 2). The input signal of desired symmetry is formed by generating, at the left of the calculated cell, forced oscillations of boundary elements of the form:

$$u_i(1, i, t) = u_{i,0}\sin(\omega_0(t - t_0))exp(-(t - t_0)^2/2\sigma_t^2) \tag{8}$$

the Fourier transform of which is a Gaussian centered at frequency ω_0 with dispersion $\sigma_\omega \sim 1/\sigma_t$. On the left side of the calculated cell forms therewith a wave packet of width $\sim v\sigma_t$ propagating along the x direction (Figure 3). The propagation velocity and wavelength of the wave packet are determined from the dispersion law of the infinite waveguide (Figure 1). After the formation of the wave packet the system is left to its own devices.

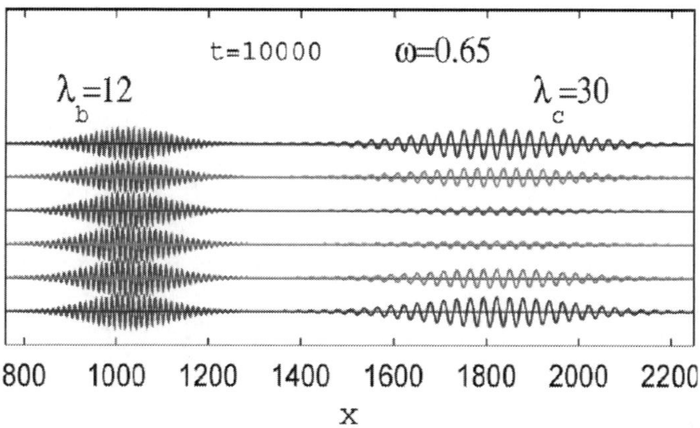

Figure 3. Wave packets of input signals of different polarization at the same instant of time.

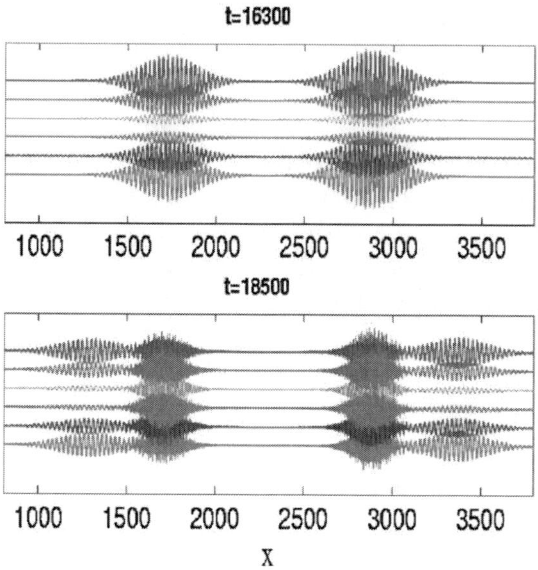

Figure 4. Wave packets of the c-wave after scattering at s- (top panel) and n- (bottom panel) defects.

To analyze the processes of reflection and transmission, the elastic energy current through detectors D_i was calculated [33]:

$$I(x_i, t) = -c_{xjkl} \int dy \partial_t u_j \partial_k u_l$$

where the integration is over the cross section b_0 at x_i.

The time integral of this expression gives the total elastic energy passed through the detector, and the ratio of the transmission (E_{tr}) /reflection (E_{ref}) to the incoming (E_{inc}) signal defines the transmission/reflection coefficient: $T = 1 - R = E_{tr}/E_{inc}$. For the given length of the main waveguide the limits on the time integral were chosen such that the signal reflected from the waveguide ends does not reach the detectors, and at the same time the condition $E_{inc} = E_{tr} + E_{ref}$ is fulfilled.

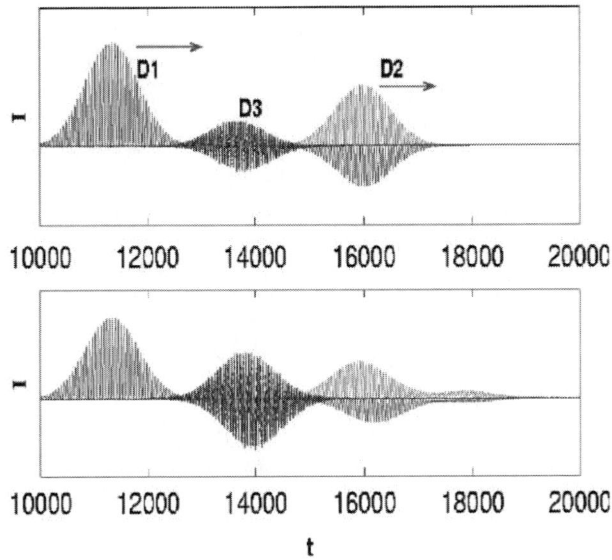

Figure 5. Elastic energy currents upon scattering of the c-wave at s- (top panel) and n- (bottom panel) defects.

3.2. Calculation Results

Consider the scattering of phonons at the bifurcation points in ZrO_2 nanowire (Figure 2). A typical picture of the displacement field of the waveguide, $(U(x, y) = x_i + Au_{xi}, y_i + Ay_{yi})$, after the scattering of a compression wave of frequency $\omega_0 = 0.725$, the transmission coefficient of which is the same $(T(\omega_0) = 0.59)$ for the symmetrical and nonsymmetrical defects, is shown in Figure 4.

As was to be expected, in the case of a nonsymmetrical defect, in the transmitted and reflected signal there are waves of both polarizations with wavelengths corresponding to the dispersion law (Figure 1). In Figure 5 are shown elastic energy currents passing through the detectors for the cases presented in Figure 4.

The frequency spectrum of the transmission coefficient $T(\omega)$ of bending and compression waves at symmetrical and nonsymmetrical defects is shown in Figure 6. In spite of certain differences, the extrema of all these dependences

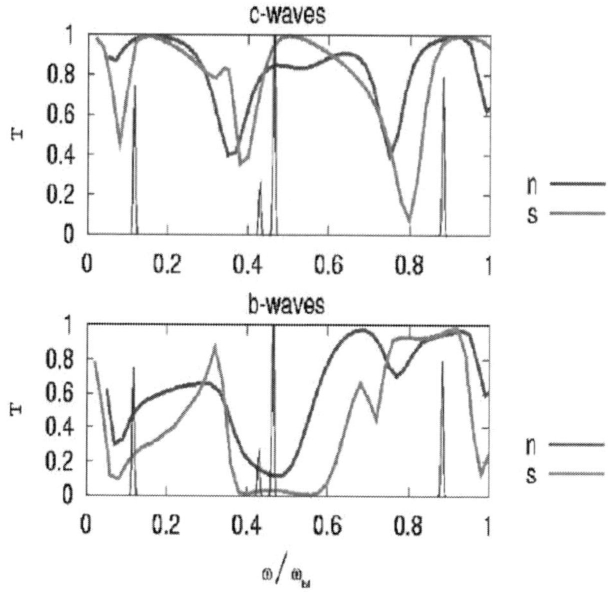

Figure 6. Transmission spectra of the c- (top panel) and b- (bottom panel) waves. The sharp peaks are the vibrational spectrum of an offshoot rigidly fixed at the base (see Figure 8).

are close together in position, which allows us to analyze the simplest case of the frequency dependence of the transmission coefficient of compression waves scattered at an n-defect. It should be noted, first of all, that on frequencies at which the maximum of the transmission coefficient is observed, the energy exchange between the waveguide and the offshoot is small, while at frequencies corresponding to the minima of the transmission coefficient there occurs intense energy exchange (Figure 7).

In order to explain this fact, the eigenfrequencies and displacement fields in these modes were found by the FDTD method for both a free offshoot (with free boundary conditions on all the surfaces) and a rigidly fixed at the base one (a column). The analysis performed showed that in interpreting the $T(\omega)$ behavior one should consider the eigenfrequencies and vibration symmetry of the column. In this case, in the spectral region under consideration there exist four eigenfrequencies of elastic vibrations of the column, three of which correspond

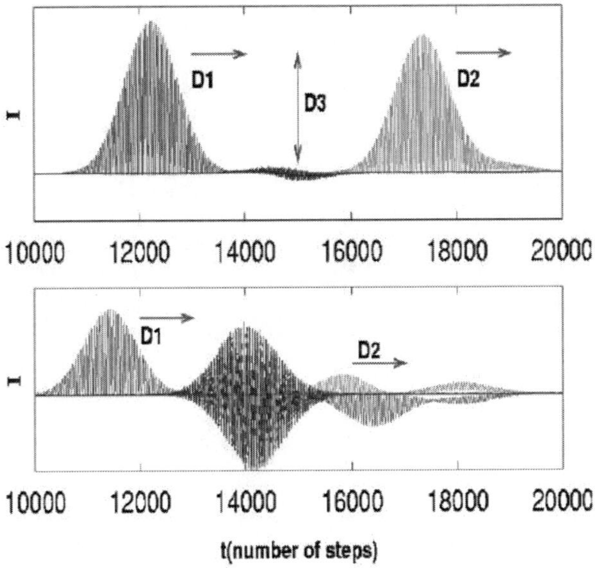

Figure 7. Elastic energy currents at frequencies $\omega = 0.9\omega_{b1}$ (top panel) and $\omega = 0.75\omega_{b1}$ (bottom panel) upon the c-wave scattering at the n-defect.

to the bending and one to the compression vibrations (Figure 8).

The n-defect vibration symmetry on frequencies at which the $T(\omega)$ minima are observed for a c-wave, coincides with the symmetry of the corresponding bending vibrations of the column. By way of example, in Figure 9 are shown the instantaneous displacement fields of the offshoot and the adjacent area of the main waveguide for one vibration period, at regular intervals equal to a quarter-period at the considered frequency $\omega = 0.75\omega_{b1}$.

Correlation between the $T(\omega)$ minima positions and the eigenfrequencies of the fixed offshoot (the gray line in Figure 6) is clearly visible, although the minima are shifted into the region of lower frequencies owing to the fact that the column base is not completely rigid. Since the c-waves do not excite compression vibrations in the defect, only three pronounced minima are observed in the frequency dependence of the transmission coefficient of these phonons, which considerably distinguishes the transmission spectrum of c-waves from that of b-waves, for which in the spectral range $0.4\omega_{b1} < \omega < 0.6\omega_{b1}$ one can observe

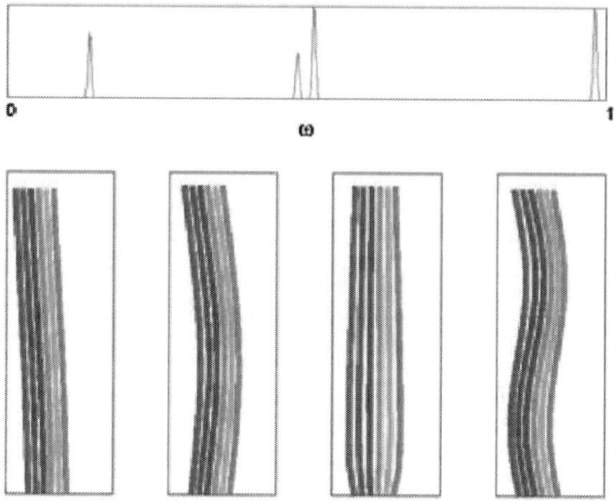

Figure 8. The vibrational spectrum (top panel) and vibration symmetry (bottom panel) at the corresponding eigenfrequencies of an offshoot rigidly fixed at the base. The lines correspond to different layers of the specimen.

a wide minimum for the nonsymmetrical defect and a forbidden band for the symmetrical one.

This result can be easily explained. It is known that the spectrum of transmission through the defect region of a homogeneous system correlates with the phonon spectrum of the periodic system including this region as its periodic element [16,34], and the correlation increases with the number of identical scatterers in the defect region (see Figure 13. from [16]). In this work we had no intention to confront the transmission and phonon spectra. However it is not difficult to understand that the peculiarities of the phonon spectrum of a quasi-one-dimensional periodic system with columns should be connected with the resonance scattering of phonons, propagating in the main waveguide, at frequencies close to the column eigenfrequencies, somewhat renormalized by the column interaction with the main waveguide (see, for example, [35,36]).

To explain this effect, we consider the defect vibrations in more detail. In Figure 10 the instantaneous displacement fields of a symmetrical offshoot and the adjacent waveguide area taken at regular intervals equal to one quarter-period are shown at the transmission frequency $\omega = 0.65\omega_{b1}$ for a c-wave (top panel) and a b-wave (bottom panel). The frequency value was chosen somewhat

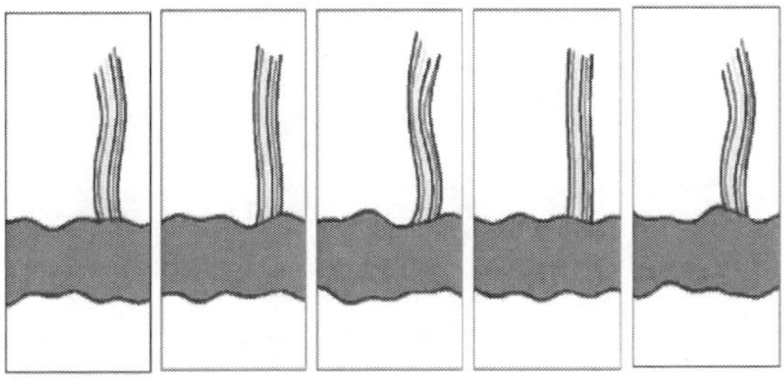

Figure 9. Instantaneous displacement fields of the n-defect and the adjacent area of the main waveguide (shaded) on transmission of a c-wave at instants of time $t = t_0 + T/4n$, ($n = 0, ...4$; T is the vibration period at $\omega = 0.75\omega_{b1}$).

above the forbidden band in order for the transmitted bending wave to have a finite amplitude. It follows from the figures that the symmetry of the defect vibrations drastically differs for the phonons of different polarization. Contrary to the case of the c-wave, the displacement field of the defect for the transmitted b-wave represents a superposition of bending and compression vibrations, which is due to the fact that the half-wave length of bending vibrations of the main waveguide, ($\lambda/2 \simeq 6\Delta x$), approximately coincides with the defect cross-section.

So the extrema of the transmission spectra for the c- and b-waves are shown to be similar and connected with the bending eigenvibrations of the defect, while the difference between the spectra is due to the compression strain of the defect excited by short-length bending waves.

4. Elastic Energy Propagation in Nanowires

4.1. Model and Calculations

Consider a two-dimensional system comprising two major blocks (source and reservoir) connected together by a thin waveguide (Figure 11). We assume all the system elements to consist of the same material (zirconia), and the size of the blocks to be $B_s = B_r = B$. This geometry, typical of a number of experiments, has been often considered in theoretical papers [26,27].

The kinetic energy of elastic vibrations in the unit cell with coordinates n, m was calculated by the formula $e_{n,m} = \rho/2(v^2_{nm,x} + v^2_{nm,y})\Delta x \Delta y$, where $v_{nm,\alpha}$

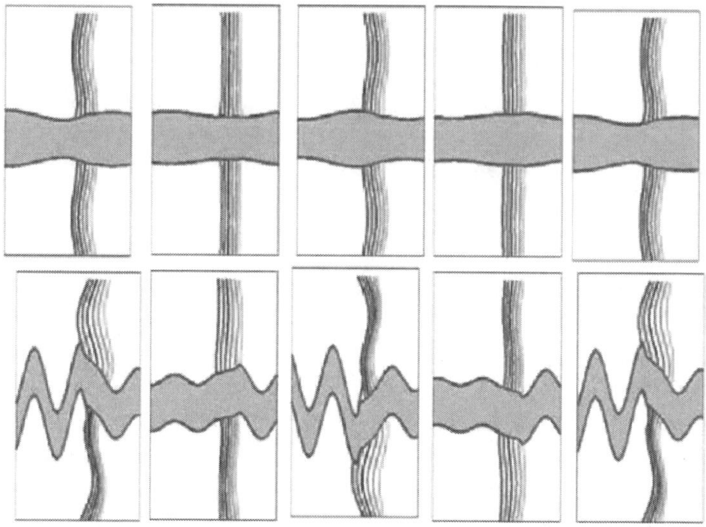

Figure 10. Instantaneous displacement fields of the s-defect and the adjacent area of the main waveguide on transmission of the c-wave (top panel) and b-wave (bottom panel) at the same instants of time as in Figure 9.

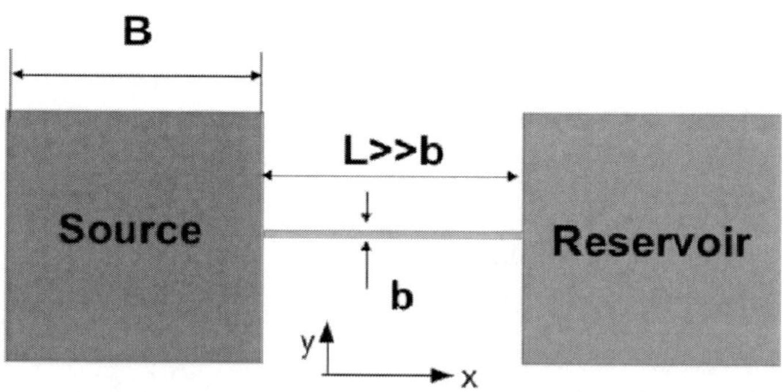

Figure 11. The calculated cell for investigation of elastic energy propagation in nanowires.

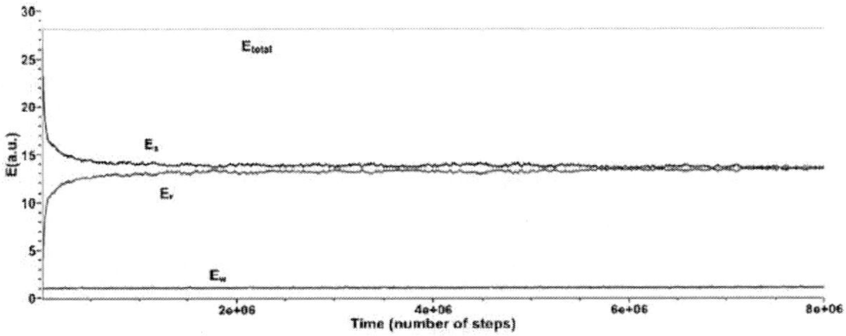

Figure 12. The time dependence of the energy of the finite system elements (B = 100, b = 5, L = 1000) upon attainment of equilibrium.

is the α-component of the velocity. The energy of the particular system elements (source, reservoir and waveguide) $E_{i=s,r,w}$ was obtained by summation of $e(n, m)$ over the indices corresponding to the given element. Besides when calculating the energy this quantity was averaged over a time interval containing about hundred time steps. Since in the 2D case the units of measurement of energy E_i, $e(n, m)$ are off-system, the corresponding results are presented in arbitrary units.

4.2. Attainment of Equilibrium in Finite Systems

Before proceeding to numerical simulation of the process of elastic energy propagation, the source should be prepared. At an instant $t = 0$, a deformation field is introduced into an isolated source with free boundaries, and then the system is left to its own devices. The system undergoes relaxation to dynamic equilibrium, in the course of which the kinetic energy is redistributed over the block volume until the equilibrium configuration of the energy field is reached. Owing to free boundary conditions the kinetic energy of surface layers is much larger than inside the block, rapidly decreasing with an increase in depth. After equilibrium is established in the source, the waveguide is opened and there occurs equalization of the energies of the source and reservoir.

The time dependence of the energies of the source, reservoir and nanowaveguide $E_i(t)$ $(i = s, r, w)$ is presented in Figure 12. The upper line on the plot corresponds to the total energy of the system $E_{total} = E_s + E_r + E_w$, which is sure to remain constant. In view of this the energy flow through

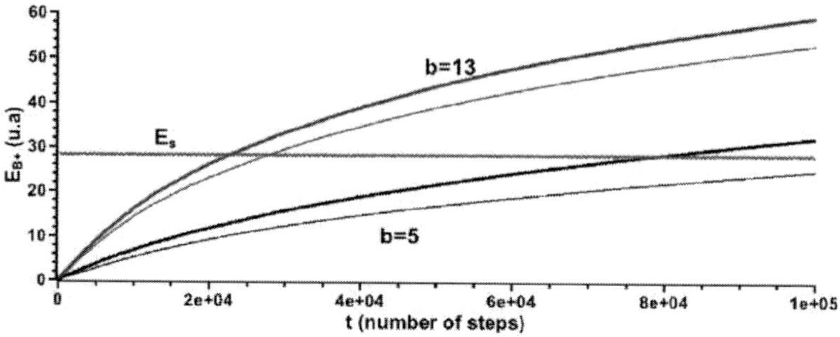

Figure 13. The time dependence of the energies of the source with B = 100 and waveguides of thickness b = 5 and b = 13 in the case of an infinite waveguide under free (thick lines) and hard (thin lines) boundary conditions.

the waveguide at instant t and coordinate x can be defined by the expression $J(x,t) = -dE_{x-}(t)/dt = dE_{x+}(t)/dt$, where $E_{x\pm}(t)$ is the system energy to the right (+) or the left (−) of coordinate x. Once a constant (maximum) energy value of the waveguide is attained, the energy flow through the waveguide no longer depends on the coordinate x and is written as $J(t) = -dE_s(t)/dt = dE_r(t)/dt$. The representation of the system time evolution in terms of $E_i(t)$ turned out to be much more descriptive than in terms of $J(x,t)$, so this representation is used in our work. The finiteness of the system size results in nonmonotonicity of the dependence $E_{s,r}(t)$ characterized by oscillations, the amplitude of which decreases with time. These oscillations induced by elastic wave backscattering from the end walls of the calculated cell become less pronounced with increasing waveguide length L. To take no account of these oscillations we consider the energy withdrawal from the source into an infinite waveguide.

4.3. Energy Propagation in Infinite Waveguide

Consider the process of kinetic energy withdrawal from the source into an infinite waveguide at a constant source energy. (In numeric calculations by "infinite" waveguide is meant the situation when the time of observation satisfies the condition $t < L/c_l$ at which the energy flow does not reach the reservoir). The source energy is kept constant by speed scaling. For this purpose the energy

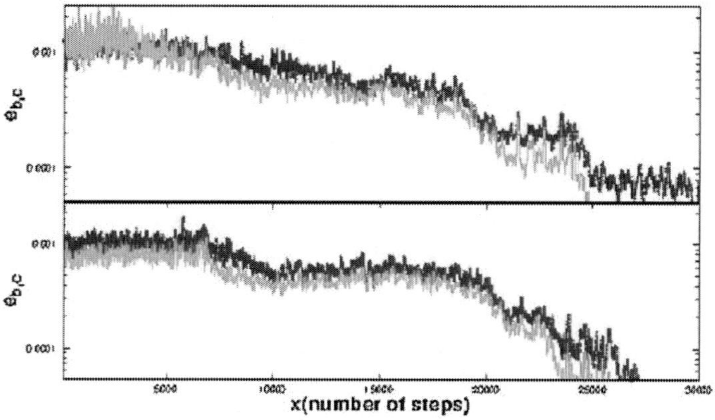

Figure 14. Energy distribution along the x-axis for compression (black) and bending (gray) vibrations under free (top panel) and hard (bottom panel) boundary conditions on the waveguide surface.

$E_s(t)$ averaged over a small time interval (about 100 time steps) is compared with a specified energy E_{0s}. In the case of deviation of the current energy value from the

preset one, the speed in all unit cells of the block is renormalized by the same factor $v' = v(1 + \alpha sgn(E_{0s} - E(t)))$, α being an adjustable parameter whose value in the calculation did not exceed 10^{-3}. In this case, for $x = B$ the flow will be expressed in terms of the time derivative of the input energy E_{in} introduced into the source to keep its energy constant, this derivative being readily calculated. Figure 13 presents the time dependence of the energies of the source and waveguide in the case of infinite waveguide for two b values and different boundary conditions on the waveguide surface: free (thick lines) and hard (thin lines). At hard conditions the surface layers are fixed, and only the section of movable layers is taken as b. Owing to the system symmetry axis, analysis is conveniently performed by passing to symmetrized coordinates, i.e. symmetrical and asymmetrical displacements with respect to the waveguide axis which describe compression (c) and bending (b) waves, respectively. Since the displacement of the element n, m $u_{nm} = \mathbf{u}_{nm,c} + \mathbf{u}_{nm,b}$, and that of its conjugate with respect to the symmetry axis $n, \overline{m}$ $\mathbf{u}_{n\overline{m}} = \mathbf{u}_{nm,c} - \mathbf{u}_{nm,b}$, the contributions to the displacement from compression (bending) waves are easily calculated,

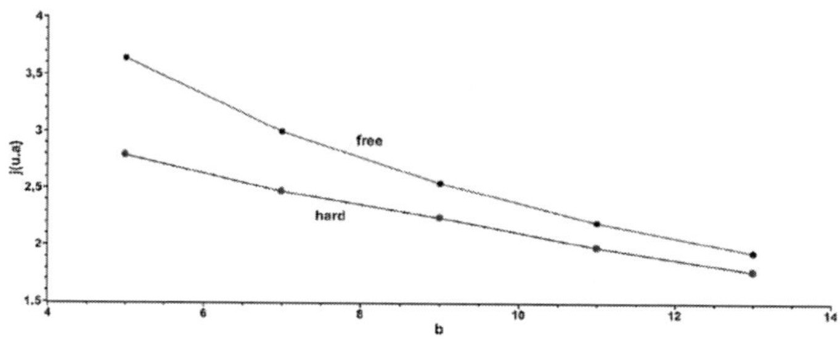

Figure 15. The dependence of the flow density $j = J/b$ on the waveguide thickness for the two boundary conditions.

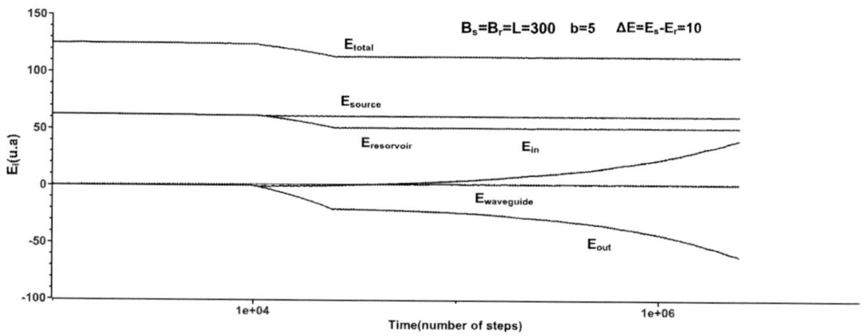

Figure 16. Time evolution of the system energy in the process of attainment of the steady-state flow (see the text).

which makes it possible to compute the corresponding kinetic energies. In Figure 14 is shown the energy distribution along the x axis, $E_i(n) = \sum_m e_i(n, m)$ $(i = c, b)$, at an instant $t = 10^5 \Delta t$. For both boundary conditions the front of the energy flow is formed by compression vibrations, which is due to the fact that the group velocity of compression waves (v_c) is greater than that of bending waves (v_b) (for low-frequency acoustic waves $v_c = c_l$, v_b [21]). The shape of the curves becomes different as the source is approached. At hard boundary conditions the bending modes are suppressed, and their contribution to the total energy is smaller than that from compression vibrations throughout the length of the waveguide. Free surface does not preclude the formation of bending vibrations, hence away from the flow front (near the source) the bending vibration energy becomes larger than the compression energy.

Figure 15 presents the dependence of the flow density ($j = J/b$) on the waveguide thickness for the two boundary conditions. It can be seen that the flow density in the free-boundary waveguide is greater than in the case of hard boundary conditions. The fact that the flow density increases with decreasing b implies that the surface layers greatly affect the process of energy propagation. The energy flow was calculated from the slope of lines approximating the dependences $E_{B+}(t)$ for times greater than $8 * 10^4$ time steps, although it is seen from Figure 13 that the flow into the infinite waveguide (at least in the calculated cell considered ($L = 40000\Delta x$) does not attain the steady-state value at which $E_{B+}(t) \sim t$. Therefore we shall create a steady-state flow in a finite system.

4.4. Steady-state Energy Flow in Finite System

To create a steady-state energy flow between the source and the reservoir the following scenario was realized. At the beginning of the process the energies of the reservoir and the source were assumed to be equal. Thereafter the reservoir energy was lowered to the preset value of E_r by withdrawing energy from the reservoir, ($E_{out} < 0$), while the source energy was kept constant by injecting additional energy $E_{in} > 0$ into the source (Figure 16). Once the dynamic equilibrium is established, these energies become equal in magnitude, and their dependences $E_{in,out}(t)$ are linear, with a slope corresponding to the steady-state flow value. Figure 17 displays the linear sections of the dependence ($t > 2 * 10^6 \Delta t$) for various energy differences between the source and the reservoir. The flow, i.e., the slope of the lines, increases directly with ΔE, which is natural. However no dependence of flow density on the waveguide thickness is observed; the inset presents the time dependence of E_{in}/b for various b.

To elucidate the situation, let us compare the layerwise energy profiles realized in the finite system subsequent to establishment of the steady-state flow (top panel in Figure 18) with the infinite-waveguide case (bottom panel). The distribution of kinetic energy along the x-direction in these cases is shown in Figure 18 for the central (core) layer of the system and the layer coinciding with the waveguide surface layer. In both cases, within the source ($x \leq B$) the energies of these layers coincide, while in the waveguide the energy of free surface layers is greater than in the central ones, as could be expected. The principal difference between the infinite and the finite waveguide is that in the former case the energy in all the waveguide layers is smaller than in the corresponding layers

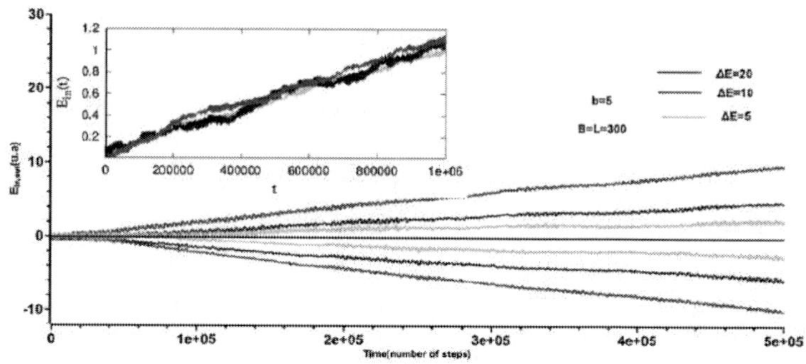

Figure 17. Time dependence of $E_{in}(t) > 0$, $E_{out}(t) < 0$ after the energy-flow stabilization at different $\Delta E = E_s E_r$. The inset shows $E_{in}(t)/b$ for different values of b. The initial points of all the curves coincide with the origin.

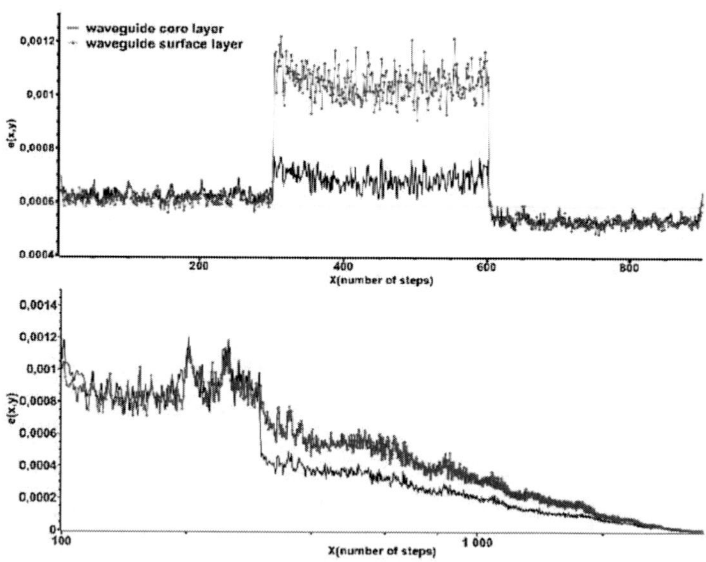

Figure 18. Energy distribution along the x-axis for the central (core) layer and that coinciding with the waveguide surface layer in the finite (top panel) and infinite (bottom panel) waveguides.

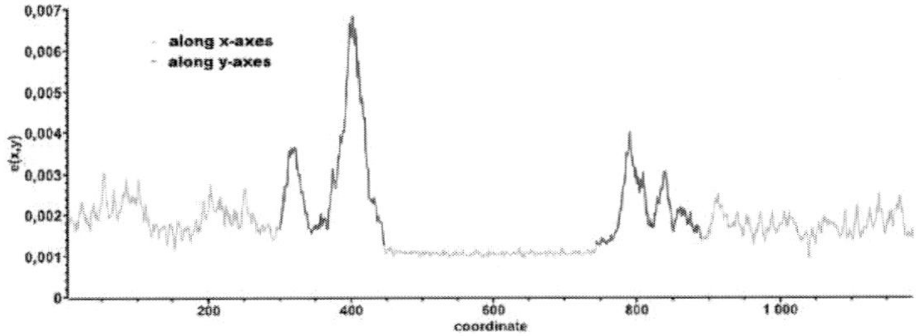

Figure 19. Energy distribution along the system surface layers from the top left to the top right point of the calculated system (see Figure 11).

of the block, which allows energy to be withdrawn from the source interior into the waveguide directly along the x axis. In the latter case the energy in all the waveguide layers is much larger than in the corresponding layers of the blocks, i.e. owing to the surface effects an energy barrier is formed which interferes with direct energy flow from the source interior into the waveguide along the x axis. Thus the energy transfer from the source into the finite waveguide occurs through the surface layers. This follows from the energy distribution along the surface layers of the blocks and waveguide shown in Figure 19. When transmitted from the waveguide, energy is propagating along the waveguide axis into the center of the reservoir where it finally takes on a configuration almost identical to the equilibrium configuration for the given block energy.

5. Conclusion

The FDTD method was used to investigate the phonon scattering at geometric imperfections of two-dimensional nanowires and to study the elastic energy propagation in mesoscopic systems of complex geometry. The transmission/reflection spectra of the phonons of different polarization (bending and compression phonons) were calculated for two kinds of defects (retaining and not retaining the waveguide symmetry). The extrema of the c- and b-wave transmission spectra were shown to be similar and connected with the bending eigenvibrations of the defect, while the difference between the spectra is due to the compression strain of the defect excited by short-length bending waves. Also, the energy field realized in the system in the process of formation and propagation of energy flow was calculated. It was shown that in mesoscopic systems the field configuration is determined by the sample geometry and boundary conditions. Based on the calculations and analysis performed the conclusion was drawn that at steady-state regime of energy propagation between two massive blocks through a small-section waveguide with free surface, energy propagates in a narrow surface layer.

This work is supported by the PPTB RAS under Grant No. 12-P-12-2010.

References

[1] Tighe T.S., Worlock J.M., Roukes M.L. *Appl. Phys.Lett.* 70 2687-2689 (1997).

[2] L. Rego and G. Kirczenow, *Phys. Rev. Lett.* 81, 232 (1998).

[3] D. Angelescu, M. Cross, and M. Roukes, *Superlattices Microstruct.* 23, 673 (1998).

[4] K. Schwab, E. Henriksen, J. Worlock, and M.L. Roukes, *Nature* (London) 404, 974 (2000).

[5] A. I. Boukai, Y. Bunimovich, J. Tahir-Kheli, J.-K. Yu, W. A. Goddard, and J. R. Heath, *Nature* (London) 451, 168 (2008).

[6] M. Maldovan. *Phys.Rev.Lett* 110, 025902 (2013).

[7] M. Maldovan. *Nature* 503, 209 (2013).

[8] J.J.Yao, W.Q. Huang, G.F. Huang, Y. Chen et al, *Current Appl. Phys.* 12 437 (2012).

[9] E.H. El Boudouti, B. Djafari-Rouhani, E.M. Khourdifi, L. Dobrzynski, *Phys. Rev. B* 48, 10987 (1993).

[10] E. H. El Boudouti, B. Djafari-Rouhani, A. Akjouj, and L. Dobrzynski, *Surf. Sci. Rep.* 64, 471 (2009).

[11] F. Garcia-Moliner, V.R. Velasco, Theory of single and multiple interfaces, in: The Method of Surface Green Function Matching, *World Scientific,* Singapore, 1992.

[12] E.H. El Boudouti, , B. Djafari-Rouhani, A. Akjouj, L. Dobrzynski, *Surf. Sci. Rep.* 64 (2009) 471.

[13] J. Sapriel, B. Djafari-Rouhani, *Surf. Sci. Rep.* 10 (1989) 189.

[14] S. Tamura, *Phys. Rev. B* 39 (1989) 1261.

[15] S. Mizuno, S. Tamura, *Phys. Rev. B* 45 (1992) 13423.

[16] X. F. Wang, M. S. Kushwaha, and P. Vasilopoulos, *Phys. Rev. B* 65, (2001) 035107.

[17] Yan Pennec, Jerome O. Vasseur, Bahram Djafari-Rouhani, Leonard Dobrzynski, Pierre A. Deymier, *Surf. Sci. Rep.* 65 (2010) 229.

[18] Vincent Laude, Abdelkrim Khelif, Sarah Benchabane, and Mikael Wilm, *Phys. Rev. B* 71, (2005) 045107.

[19] A. Taflove, The Finite Difference Time Domain Method, *Artech*, Boston, 1998.

[20] Peter Moczo, Johan O. A. Robertsson, and Leo Eisner *Advances in Geophysics,* 48, 421, (2007).

[21] L.X. Zhang , and T.Seideman, Phys. Rev.B 82, 155117 (2010).

[22] J.-H. Sun and T.-T. Wu, *Phys. Rev. B* 76, 104304 (2007).

[23] E.I.Salamatov, *Phys. Status Solidi B* 246, 92 (2009).

[24] E.I.Salamatov, *Phys. Status Solidi B* 249, 113 (2012).

[25] E.I.Salamatov, *Phys. Status Solidi B* 250, 1932 (2013).

[26] M. C. Cross and R. Lifshitz, *Phys. Rev. B* 64, 085324 (2001).

[27] M. P. Blencowe, *Phys. Rev. B* 59, 4993 (1999).

[28] G. Mur, *IEEE Trans. Electromagn. Compat.* EMC-23, 377 (1981).

[29] M. Sato, *Acoust. Sci. Technol.* 28, 1 (2007).

[30] T. Ohminato and B. A. Chouet, *Bull. Seism. Soc. Am.* 87, 494 (1997).

[31] Y. Tanaka, Y. Tomoyasu, and S. Tamuraet, *Phys. Rev. B* 62, 7387 (2000).

[32] S. Fabris, A. T. Paxton, and M. W. Finnis, *Phys. Rev. B* 61, 6617 (2000).

[33] M. P. Blencowe, *Phys. Rev.B* 59, 4992 (1999).

[34] E.I.Salamatov, *phys. stat. sol. (b),* 246, 92 (2009).

[35] A. P. Zhernov, E. I. Salamatov, and E. P. Chulkin, *phys. stat. sol. (b)* 165, 355 (1991).

[36] A. P. Zhernov, E. I. Salamatov, and E. P. Chulkin, *phys. stat. sol. (b)* 168, 81 (1991).

INDEX

A

B

C

D

E

F

G

H

I

O

P

T

U

V

W

X

Z